Living With Wildfires

Prevention, Preparation, and Recovery

by Janet C. Arrowood

"It does not do to leave a live dragon out of your calculations,
if you live near him."

— From *The Hobbit* by J.R.R. Tolkien

PLEASE READ

This book, *Living With Wildfires: Prevention, Preparation, and Recovery,* is intended to provide general information with regard to the subject matter covered. It is not meant to provide legal opinions or offer professional advice, nor to serve as a substitute for advice by licensed, legal or other professionals. This book is sold with the understanding that Bradford Publishing Company and the author, by virtue of its publication, are not engaged in rendering legal or other professional services.

Bradford Publishing Company and the author do not warrant that the information herein is complete or accurate, and do not assume and hereby disclaim any liability to any person for any loss or damage caused by errors, inaccuracies or omissions, or usage of this book.

Laws, and interpretations of those laws, change frequently and the subject matter of this book contains important legal consequences. It is the responsibility of the user of this book to know if the information contained in it is applicable to his situation, and if necessary, to consult legal, tax, or other counsel.

Library of Congress Cataloging-in-Publication Data

Arrowood, Janet C.
  Living with wildfires : prevention, preparation, and recovery / by Janet C. Arrowood.
      p. cm.
Includes bibliographical references and index.
  ISBN 1-883726-89-1 (alk. paper)
  1. Dwellings--Fires and fire prevention. 2. Wildfires. I. Title.

TH9445.D9A77 2003
628.9'22--dc21

                                      2003007328

DEDICATION

To all wildland firefighters everywhere,
and to the people who support them.

A NOTE FROM THE PUBLISHER

Not all wildland fires are "bad" fires. In fact, fire is an important natural agent of change, playing a vital role in maintaining healthy ecosystems. Recognizing this fact, forestry scientists and wildland managers have, for the past several decades, implemented plans that allow certain naturally occurring wildland fires to burn, as long as they do not threaten lives, property, or natural resources. These natural fires are fought only if they escape predetermined boundaries or if current environmental conditions are deemed too hazardous to allow the fire to continue. In addition, modern wildland management plans include "prescribed" fires that are intentionally set by wildland managers under controlled conditions and allowed to burn within confined areas. Both types of fires reintroduce the beneficial effects of fire on an ecosystem, not the least of which is reducing the hazard of catastrophic uncontrolled fires caused by the excessive buildup of natural fuels.

This book is about those catastrophic uncontrolled "bad" fires, commonly called *wildfires*. These are wildland fires that *do* threaten life, property, or natural resources—and they are on the rise. As more people are choosing to live in the "wildland/urban interface," homes are being built in and near forests and other areas where wildland fires naturally occur. And as more people are using wildland areas for recreational purposes, people are causing wildfires. Whether through discarded cigarettes, campfires, trash burning, arson, arced power lines, sparks from equipment, even by sparks from passing trains, or other careless means, wildfires are often ignited. These are the unplanned, uncontrolled, and unnecessary fires that could be most easily prevented.

Even Smokey Bear has updated his message:
Only YOU Can Prevent Wildfires.[1]

§

We have made every reasonable effort to insure that the information presented in this book is as accurate and up-to-date as possible by double checking the facts and figures it contains through additional research during the editing process. We have also had the manuscript thoroughly reviewed by experts in the fields of forestry, firefighting, and insurance and have incorporated their corrections along with many of their suggestions for improving the quality and usefulness of the book. However, it would be presumptuous of us to think the text is perfect. Therefore, we encourage our readers—experts and average citizens alike—to let us know what you think by sending your comments, corrections, and recommendations to: Editorial Department, Bradford Publishing Company, 1743 Wazee Street, Denver, Colorado 80202. To the extent possible, we will incorporate your input in future editions of the book.

— Bradford Publishing Company
April 2003

---

[1] See http://www.smokeybear.com. This Note is based on information provided on that website. Smokey Bear's image is protected by U.S. law and is administered by the USDA Forest Service, the National Association of State Foresters, and the Ad Council.

CONTENTS

FOREWORD

By Justin Dombrowski

Fire professionals are experiencing wildfire intensities of record levels in recent years, especially in the wildland/urban interface. Firefighters have lost their lives, hundreds of homes have been destroyed, and critical natural resources are suffering. Even though over 95% of all wildfires are kept small and quickly extinguished, hundreds of fires grow to thousands of acres and cause tremendous damage. There is no simple wildfire anymore, as most fires today threaten or are heavily involved in the wildland/urban interface.

The fire risk is growing as more people move into wildland/urban interface areas and visit our public lands to get away from the cities—but without the support and protection cities can provide. Firefighters and emergency responders are asked to do more, responding to a wide array of emergencies with shrinking resources. Many firefighters in mountain and rural areas are volunteers, with limited time to help. With economic and budget problems around the country, many responders have less equipment and resources and fewer firefighters are trying to respond to more fires and other threats with limited time and capabilities.

Firefighters make safety their number one priority—for themselves, their crew, and the public. This priority is simple, but the many hazards and risks involved in firefighting take training and experience to overcome. With the number and intensity of recent wildfires, limited firefighting resources, and increased development in wildland/urban interface areas, homeowners cannot realistically expect a fire engine at every house or rely solely on emergency responders to save the day. People need to take personal responsibility and make their own preparations before the next fire.

This valuable book provides important details to prepare your family and your property before a fire even starts, what to do during a fire event, and the issues you are likely to face after a fire has burned through your neighborhood. It can help you understand how to help firefighters protect themselves and provide them an opportunity to protect your family and your home. There are no guarantees when dealing with wildfires, there is no way to be 100% safe, and this book will not solve all your problems. It is, however, an important tool to help you increase the safety of your family and the survivability of your home.

As you read this book, look around your house. What pictures and memories would be lost forever if a fire were to destroy your home? What is important to you and what precious items would be gone if a fire came through tomorrow? Much of what you have in your life is irreplaceable, but there is much you can do to help protect what you own and what you care about. *Living With Wildfires* is full of tips, checklists, and information about how to prevent these losses.

Preventing fires from happening in the first place is the cheapest and most effective measure land managers and homeowners can take. Fighting fires can cost millions of dollars for a single fire event. There are extensive costs for air resources and to mobilize and support hundreds of firefighters with all the necessary equipment to do the job. Afterward, rehabilitation efforts to stabilize and protect the land can cost millions more than the suppression costs. The fire prevention steps outlined in this book are far less expensive and can provide added benefits to wildlife and the native vegetation.

Fire is natural and necessary to ecosystems, but many fires go beyond what nature intended. Thousands of fires are started by people every year, and we are experiencing more intense fires as the build-up of fuels increases. Fires are burning faster and more aggressively and can change direction and intensity in seconds when the wind takes hold. Grassfires can have flames over 30 feet high, spread over 1,000 feet a minute, and have temperatures above 800 degrees Fahrenheit. Forest fires with 150-foot-high flames can spread above treetops, deposit burning embers over a mile in front of the fire, and burn at 2,000 degrees Fahrenheit. People understand that firefighters cannot stop a hurricane or tornado, and they need to understand that some wildfires cannot be stopped no matter how many fire engines we have.

Historically, there have been single years in which wildland fires have consumed more than 20 million acres. In recent years, fires have consumed less acreage but with greater devastation because of the increased threat to life, property, and critical natural resources. Wildland fires occur in most parts of the world, but the increased threats and intensities have reached record levels. Suffering the impact of a wildfire is not a question of "if" but a matter of "when."

The wildland/urban interface is a wonderful place to live and play, but it holds a reality you must face: The wildfire problem is not going away. The need to have an educated and proactive community has never been greater. You can reduce many of the potential risks by following the details outlined in this book. It provides not only important prevention information but also what to do when a fire is actually approaching, and sadly, it covers the critical details you'll need to know and follow if a fire has damaged or destroyed your home.

It is clear Janet has gone to great lengths to research the information provided here. She has seen the impact of wildfire up close and personal, has learned many important lessons, and offers this valuable knowledge and in-depth research to help you on your way. She has created an important tool. Now you must fill your toolbox and turn your education into action.

This is a call to help all firefighters and yourself. Firefighters will continue to sacrifice their time and energy—and in some cases their lives—to protect the lives of others, as well as the homes and vital natural resources. Show them you've done your best to help them in their mission by preparing your family and property to face the threat of wildfires. This book is an important step in the right direction. Educate yourself and your community, and move forward knowing that you will make a difference.

— Justin Dombrowski[2]
April 2003

---

[2] Justin Dombrowski is the Wildland Fire Management Officer for the Boulder Fire Department and the Wildland Fire Coordinator for the Boulder Rural Fire Department in Colorado.

ACKNOWLEDGEMENTS

The author and the publisher would like to express our gratitude to Justin A. Dombrowski, Wildland Fire Management Officer for the Boulder, Colorado Fire Department, for his valuable insight and contributions to this book, and for his dedication to educating the public. We also wish to thank Michael G. Apicello, Public Affairs Officer for the National Interagency Fire Center, for his exceptional patience and support in helping us achieve our goal of enlightening residents of the wildland/urban interface with accurate information on the subject of wildland fires. Our thanks go to Jeannette Hartog and Karen Curtiss with the USDA Forest Service for their helpful evaluation of the manuscript and thoughtful suggestions. We also appreciate the meticulous review of the Insurance chapter provided by Michele S. Steinberg, Firewise Communities Support Manager for the National Fire Protection Association (NFPA), and the input we received on the Prevention chapter from James Smalley, Manager of Wildland Fire Protection for NFPA.

We wish to credit Frank Dennis, Wildfire Hazard Mitigation Coordinator with the Colorado State Forest Service for sharing vital information on creating wildfire-defensible zones and fire-resistant landscaping. Thanks also go to the High Country News, Colorado State University Cooperative Extension, Firewise, and FEMA for generously granting permission to reprint a variety of helpful materials. Additionally, we are extremely grateful to Candace Boyle, Lois Trimble, John and Cindia Hogan, and Kirk Hanes for sharing their personal wildfire experiences.

Introduction

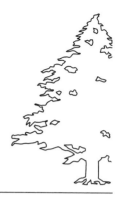

Millions of people around the world live in fire-adapted ecosystems. Many of these people are unaware that fire will ultimately return to these areas and that many of their homes will be at risk. As more people seek to escape densely populated cities and suburbs, they are building homes in rustic wooded areas near open space and public lands, described by wildland fire experts as the "wildland/urban interface." After decades of human habitation, these areas typically contain a dense mixture of vegetation, such as grasses, weeds, shrubs, and trees, as well as homes and other structures—much of which can burn and become fuel for naturally occurring and human-caused fires.

During the fire seasons of at least the past decade, images of ferocious fires and the destruction they cause have appeared almost daily in newspapers and on television newscasts, generating a high degree of public interest and concern. The 2002 fire season was one of the most severe in recent memory: over seven million acres of land in the United States burned, along with hundreds of homes and thousands of other structures located in the wildland/urban interface. Canada has also suffered significantly. And Australia was experiencing devastating wildland fires as their fire season swung into 2003.

For Arizona, Colorado, New Mexico, and Oregon, 2002 was their worst fire season ever. In ten states over 100,000 acres burned, including over 900,000 acres in Colorado, one million acres in Oregon, and two-and-a-quarter million acres in Alaska. Like a number of other populated areas confronted with menacing fires, the towns of Deadwood, South Dakota, and Show Low, Arizona, were both completely evacuated for days until the fire danger passed. Sadly, more than 20 wildland firefighters lost their lives in fire-related incidents.

This book is about the truly catastrophic uncontrolled variety of wildland fires that burn through wildland/urban interface areas threatening life, property, or natural resources. These fires are commonly referred to as

"wildfires." More importantly, this book is about the people who live in
the wildland/urban interface and what they can do to reduce the threat of
wildfires to their homes and families, what they should do if the wildfire
danger is imminent, and what they might do to pick up the pieces if they
suffer a loss due to wildfire.

Many of the words and expressions used by wildland firefighters have
specific meanings, and there are many phrases unique to the firefighting
world. A Glossary is included in the back of this book to help you understand this wildland firefighting terminology. Also included are a
Bibliography and a Webliography, listing resources for additional information on wildfires, as well as four types of evacuation checklists to help
you prepare.

WILDFIRES, NATURE, AND PEOPLE: AN OVERVIEW

How bad was the 2002 fire season? How serious is the threat of future
fires? The following chart appears on the website of the National
Interagency Fire Center in Boise, Idaho, and illustrates the number of fires
and acres burned, by state, in that one year.[3]

Total Number of Wildland Fires and Acres in the U.S. January 1 to November 15, 2002, by State

State	Fires	Acres
Alaska	547	2,267,380
Alabama	3,005	39,221
Arkansas	1,331	14,405
Arizona	2,860	650,466
California	7,622	491,333
Colorado	3,072	915,291
Connecticut	101	184
Delaware	30	1,659
Florida	2,495	50,304
Georgia	7,179	159,951
Hawaii	1	3,660
Iowa	5	1,045
Idaho	1,483	84,854
Illinois	27	94

---

[3] See www.nifc.gov/fireinfo/nfnmap.html.

Indiana	578	1,684
Kansas	24	2,659
Kentucky	1,005	25,092
Louisiana	1,120	22,566
Massachusetts	3,018	2,611
Maryland	726	4,825
Maine	662	784
Michigan	277	966
Minnesota	2,013	62,935
Missouri	89	3,096
Mississippi	1,006	16,601
Montana	1,412	111,819
North Carolina	5,417	34,320
North Dakota	811	58,686
Nebraska	45	427
New Hampshire	424	168
New Jersey	1,760	6,053
New Mexico	1,822	325,458
Nevada	734	85,099
New York	336	2,044
Ohio	577	3,969
Oklahoma	1,006	29,161
Oregon	2,631	1,010,844
Pennsylvania	529	1,961
Puerto Rico	2	4
Rhode Island	128	213
South Carolina	3,567	30,943
South Dakota	639	30,679
Tennessee	1,469	14,993
Texas	1,104	29,403
Utah	1,226	265,337
Virginia	1,699	23,294
Vermont	128	214
Washington	1,281	88,863
Wisconsin	800	1,611
West Virginia	814	8,681
Wyoming	523	124,823
Grand Totals	71,160	7,112,733

According to a document titled "Healthy Forests: An Initiative for Wildfire Prevention and Stronger Communities" issued by the White

House on August 22, 2002, noting that the 2002 fire season was already by that date one of the worst in modern history:

> "Already more than 5.9 million acres of public and private land have burned this year, an area the size of New Hampshire and more than twice the average annual acreage, with more than a month of fire season remaining. Fires have burned 500,000 acres more than they had at this time during the record-setting 2000 fire season.

> "Hundreds of communities have been affected by these wild-fires. Tens of thousands of people have been evacuated from their homes, and thousands of structures have been destroyed. With more people living near forests and rangelands, it is becoming increasingly difficult to protect people and their homes. Land managers must do more to address the underlying causes of these fires.

> * * *

> "America's public lands have undergone radical changes during the last century due to the suppression of fires and a lack of active forest and rangeland management. Frequent, low inten-sity fires play an important role in healthy forest and rangeland ecosystems, maintaining natural plant conditions and reducing the buildup of fuels. Natural, low-intensity fires burn smaller trees and undergrowth while leaving large trees generally intact. Natural fires also maintain natural plant succession cycles, pre-venting the spread of invasive plant species in forests and range-lands. This produces forests that are open and resistant to dis-ease, drought, and severe wildfires.

> "Today, the forests and rangelands of the West have become unnaturally dense, and ecosystem health has suffered signifi-cantly. When coupled with seasonal droughts, these unhealthy forests, overloaded with fuels, are vulnerable to unnaturally severe wildfires. Currently, 190 million acres of public land are at increased risk of catastrophic wildfires."[4]

---

[4] From "Healthy Forests: An Initiative for Wildfire Prevention and Stronger Communities" (August 22, 2002), available at http://www.whitehouse.gov/infocus/healthyforests/Healthy_Forests_v2.pdf.

The Role of Fire in Forest Ecosystems

Throughout the first three-quarters of the 20th Century, complete fire suppression—at almost any cost—was the socially acceptable firefighting approach used by most foresters and wildland managers in the West. During that era, however, these managers came to realize that fire, not just frequent, low-intensity fire, was essential in ecosystems to maintain healthy forests and other wildlands. Nevertheless, the only kinds of fire allowed to burn, as a rule, were the low-intensity ones, because managers were wary of fires that "crowned out," spreading ferociously from tree-top to tree-top and threatening anything that could burn for miles.

Fires are essential for many ecosystems, maintaining a desirable balance and mix of plants and animals and reducing the build-up of deadfall and other potential fuels. Naturally occurring fires, along with "prescribed burns" (fires set and controlled by wildland managers), help to eliminate small, densely packed trees and undergrowth while leaving larger trees relatively unscathed. There is also a natural succession of plant growth, death, and restoration that depends on fire. Without it, non-native plants may take over, destroying ecosystems and worsening future fires while driving out native fire-adapted plants and wildlife. Properly managed, fires often improve the health of both wildlands and animals, helping create forests that are more resistant to drought, insects, invasion by non-native plants and animals, disease, and more severe fires.

More specifically, there is a range of ways that forests and trees are adapted to fire. Lodgepole pine, for example, is adapted as a species to fire because it needs fire for regeneration. Its adaptation is that when fires occur, large pockets of trees are consumed; but seeds stored in cones are released as a result of heating of the cones, and these seeds produce massive regeneration after the fire. Lodgepole pine *trees* themselves are not adapted to survive fire. But without fire their correct place in the ecosystem may be lost; other species (such as Engelmann spruce and sub-alpine fir) invade Lodgepole pine forests and begin to crowd them out. Fire kills those species, and they do not regenerate readily after fire because they have no stored seeds.

Ponderosa pine trees, on the other hand, are more adapted to survive fire, and in historical forests, trees in the "overstory," i.e., the tall mature trees, often experienced a number of fires during their lifetime. Because of thick bark, relatively sparse foliage with long needles, and natural pruning of lower branches, Ponderosa pine trees often survive fire, unless it burns extreme.

The Effect of Fire on Forests

Fire affects forests differently depending on the type of trees involved and the interval of time since the last fire. Two examples, again, are Lodgepole and Ponderosa pine forests:

Lodgepole pine forests, historically, burned with nearly complete mortality across large areas, as in the 1988 Yellowstone fire. These were "active" crown fires in which fire readily spreads from one tree crown to the next. Surface fires were uncommon in these forests, and would have been lethal to most Lodgepole pine trees anyway because they do not have thick bark or protected foliage as found in Ponderosa pine.

Depending on location, Ponderosa pine forests have experienced several different kinds of fire behavior. In southwestern states and some of the west coast states, fires have generally been frequent (every 3-10 years or so) and have burned with low intensity. In part, climate and climate change produced lots of understory vegetation in wet monsoon years and provided fuel for fires to spread. Those fires created forest stands that often were open and "park like."

But in the Colorado Front Range—and probably most places in the interior West beginning at a latitude just south of Pikes Peak—monsoonal moisture from the southwest is less reliable, and fuels to support fire spread are more limited in *most* years. Accordingly, fires historically were less frequent, even as long as 40-60 years apart—though a 5-20 year cycle is more common. During the time between fires, fuels slowly accumulated in the form of dead branches and needles beneath trees, small trees creating fuel ladders, and shrub fields. So when conditions resulted in enough fuels to spread surface fire, these fires could simply burn beneath some trees, but could also kill individual or large numbers of trees. This type of fire behavior is called "mixed-severity" fire behavior.

Part of every mixed-severity fire burned just on the surface as a surface fire, not unlike the southwest pattern. But part of mixed-severity fires could burn as a "passive" crown fire in which fuels or small trees beneath larger trees caused "torching" of individual trees, or as a hotter surface fire in which lethal temperatures are reached without combustion of the crowns of the larger trees (so trees died by scorch rather than by torching).

Depending on fuel conditions and local weather patterns, torching and scorching could kill some trees and leave others, or it could kill all trees on a hillside and create an opening (especially if a large patch of shrubs developed beneath the sparse forest overstory). Only in very limited cases, such as a thicket or patch of regeneration that created a small but dense patch of trees, could an active crown fire have occurred in most of these forests. Thus in many Ponderosa pine areas, historical mixed-severity fire patterns maintained a complex landscape structure, with a lot of openings or areas with very few trees per acre dispersed across the landscape, with some other areas being somewhat more dense.

During the last century or more, nearly all Ponderosa pine forests experienced increases in forest density as a result of logging and grazing, which favored the establishment of many new seedlings, and fire suppression policies that allowed fuels to accumulate and tree densities to become too high. The result today is that, whatever the historical fire behavior pattern in Ponderosa pine (frequent low-intensity surface fire or mixed-severity fire), many forests are now dense enough to have active crown fires. Although active crown fires are appropriate for Lodgepole pine, they are less appropriate for Ponderosa pine. For that reason, current Ponderosa pine forests that have become too dense are considered to be ecologically unsustainable, while Lodgepole pine forests naturally get dense and self-thin, and they are not considered unsustainable. Thus management activities to thin and remove fuels in Ponderosa pine forests make good sense ecologically, if they mimic low-intensity cleansing fires, but such activities would not be considered good ecology in Lodgepole pine forests.

How *Should* a Forest Look?

What is considered "ideal" density for a forest today varies significantly among different types of trees, elevations, and latitudes, and according to the intended or actual use of the particular tract of land, whether for

homes, recreation, lumber, wildlife habitat, wilderness, or water quality, for example. By way of illustration, the two pictures below demonstrate the difference between "stagnated" and "ideal." In the picture on the right, there are over 1,000 trees per acre and an accumulation of natural fuels; but a healthy pine forest might have fewer than 500 trees per acre, and 200-300 is even better. The picture on the next page shows one version of what might be considered the "ideal" forest, where the trees have been thinned and the understory, groundfall, and other potential fuels have been removed.

As one forestry expert stated, "This dense condition is probably completely typical in naturally occurring Lodgepole pine forests, even though it may cause a wildfire hazard. That's simply the result of us building houses where we shouldn't. This is a good example of [a situation] when thinning might be justified by our social choices of where to build and live. But don't thin these forests and call it good ecology."

How Do Fires Burn?

"Fuel, heat and oxygen are all needed in the right combination to produce fire. Combined, they're called the 'fire triangle.' By nature, a triangle needs three sides. Take away one of the sides, and the triangle collapses. The same is true of fire. Take away any of the three components of fire—fuel, heat or oxygen—the fire collapses, meaning that it can't burn. Firefighters try to do just that—remove one of the three essential components of fire. For example, when they dig a line around a fire, fuel is removed. When water is dropped on a fire, it reduces the heat. Retardant, a thick, soupy substance, coats fuels, blocking them from oxygen. If you think of fighting fire in terms of breaking the fire triangle, then it's easier to understand the tactics of firefighters." From "This Thing Called Fire" by the National Interagency Fire Center, http://www.nifc.gov/pres_visit/whatisfire.html.

Photograph from www.wheremedia.com/nifc/index.html.

Wildland fires are caused both by nature, through lightning, and by people, through discarded cigarettes, campfires, trash burning, arson, arced power lines, sparks from equipment, even by sparks from passing trains, or other careless means. Once a wildland fire starts, it can burn unchecked for hours or days before it is noticed. Fires have a way of igniting off the beaten path and then "spreading like wildfire," to use a phrase we've heard a lot of lately. Wildland fires become dangerous and destructive wildfires when they threaten lives, property, and critical natural resources.

There are three main types or classes of fires in forested areas: surface fires, ground fires, and crown fires. These fires may occur independently or simultaneously.

Surface fires are those that burn through the undergrowth of a forest. (Surface fires can also sweep through grasslands with deadly speed, up to 1200 feet per minute, and with flames up to 30 feet.) When surface fires are low in intensity, they serve the important role of clearing out debris and deadfall on the forest floor. However, in high wind conditions, surface fires burn much hotter and can severely damage or kill trees by climbing up "ladder fuels" (see the following section) to become crown fires.

Ground fires are usually started by lightning and burn *on or below* the forest floor (even in the roots of trees). These fires damage but rarely destroy mature trees, but they do a great job clearing out densely packed seedlings and creating open meadows that form a necessary buffer area between stands of mature trees.

Crown fires are those scary fires we experienced during the massive killer fires that erupted in 2000 and 2002 across much of the West, and in other years around the country and the rest of the world. These are the fires that quickly run up the trunks of densely packed or dead trees, spreading rapidly (due to wind) from the crown of one tree to others, appearing to "jump" from treetop to treetop. The tree crowns provide both abundant fuel and ready access routes to other trees and to structures. Nevertheless, crown fires often drop out of the treetops or go out altogether when they reach areas treated with prescribed burns or thinned by other mechanical means.

Understanding Fuel Loading, the Fuel Continuum, and Ladder Fuels

The Glossary defines three critical wildfire 'fuel' terms:

- Fuel Loading
- Fuel Continuum
- Ladder Fuels

What do these terms really mean, and why are they important?

Fuel loading refers to the amount and composition of flammable materials in a given location. The process of creating a "defensible space" around your home is focused on eliminating the fuel load that might carry the fire to your home. (See Chapter 1: Prevention.) Fuel loading is of great concern in the wildland/urban interface, since the denser the stand and more flammable the fuel load, the greater the risk that once a fire is ignited it will be able to establish itself, increase in intensity, and spread. The fuel load in a given area may be all one type of material, or it may be a mixture. Both the type of material and the density of material affect the degree of flammability and fire intensity. If you live surrounded by areas with significant fuel loading, you are far more likely to be the victim of a serious, high-intensity fire than if the fuel loads are managed and/or reduced through prescribed and controlled burns, or eliminated altogether through the creation of your defensible space.

The fuel continuum is the spectrum of flammable material in a given area. How flammable the fuel continuum might be is affected by the types of fuels it contains, their structure, size, diversity, and moisture level. The fuel continuum may include:

- Fine *tinder* such as pine needles, dead leaves, paper, grasses, bark, and small twigs.
- More substantial *kindling* such as branches, bushes, small trees, mounds of trash, deadfall, diseased or damaged trees, chaparral, and slash.
- Material that can sustain a fire such as logs, larger trees, and structures.

If the fuel continuum is missing significant amounts of kindling, any fire that occurs is far more likely to be a low intensity one—the type that burns the grasses, singes the tree trunks, and burns itself out rather quickly—a surface or ground fire. If there are large amounts of kindling available, the resulting fire is far more likely to become well-established and be able to use the various types of kindling to lift itself into the larger trees and up into the crowns—this results in a high intensity crown fire.

This brings up the subject of ladder fuels. These are any series of vegetation types or other combustible materials arranged in a vertical continuum that allows a surface or ground fire to climb into the tops of taller trees. The picture on the next page gives you an idea of what ladder fuels look like. Literally, they form a series of "steps" that can give a fire lift into the trees, leading to a crown fire. Note the debris and deadfall on the

ground, the bushes and smaller trees, the low-hanging branches, and the way these potential fuels lead right to the mid-point of larger pine trees. If a fire started here, it would have a very easy time making its way up into the treetops, and becoming an intense crown fire.

Fires can normally only establish themselves if they have the right mix of fuels (tinder, kindling, and then more substantial materials) and the right climatic conditions (generally warm temperatures, low humidity, and wind). They can only continue to burn if they have fuel to consume. Wildfires establish themselves in much the same way as you start a campfire or a fire in your fireplace: they need tinder to start, kindling to get established, and then more substantial fuels to spread. If you try to light a fire with lots of tinder, little or no kindling, and big logs, your fire is likely to go out when it burns the tinder. So it is with most wildfires; if you take away their kindling fuels, they burn with less intensity. You have to add the kindling to really get most wildfires, campfires, and fireplaces going. Take away the middle component, and you generally increase the likelihood a fire will never get going in the treetops (but may still continue to burn on the ground at lower intensity). Create an effective defensible space and you do the same thing, greatly increasing the survivability of your structures.

What This Means For You

Owning a home in the "wildland/urban interface" means you live in the *in-between* area where the city, suburbs, and residential developments thin out and become more sparsely populated, or where these urban areas butt right up against the countryside, open space, or public lands. It means you have chosen to live and work surrounded by nature, with all its beauty—but also with all of its hazards, including periodic fire from both natural and human causes. Take some time to walk around your

Photograph from www.wheremedia.com/nifc/index.html.

neighborhood and hike in the surrounding wildlands. You'll probably see areas of exceptionally dense trees, shrubs, undergrowth, and grasses. Ask yourself, "What would happen if a wildfire were to burn through here? What might be done to minimize the damage?" This book will help you reduce the risk of fire damage to your property and the surrounding landscape, be safer in the natural environment, and minimize the impact on your life and family if you become a victim of wildfire.

Chapter 1: Prevention

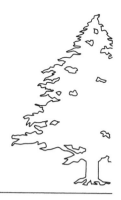

o you live in the woods, or the foothills, or on a ranch, or in some other semi-secluded area a little off the beaten path? Then you probably live in what is often referred to as the wildland/urban interface. You're not in the deep woods, but you're not in a city either. You may be in a town, but not in a truly "urban" area, and your house is a long way from the nearest firefighting resources, even a volunteer fire department. You are living in the place where the city and the country meet, hence the term "wildland/urban interface." You enjoy the beauty and seclusion of mountains, canyons, ranches, and other wildlands, as well as the convenience and relative proximity of the city. But these idyllic places are fraught with risks you may never have considered—chief among them is the likelihood of wildfires, not to mention the flash floods that often follow in their wake.

Advance planning by people who live in the wildland/urban interface—as well as by people who use recreational, wildland, and open space areas—can help protect lives, buildings, and property, and lessen the devastation of a wildfire. In this chapter, you will learn many of the ways to minimize the threat of damage to or destruction of your home, property, and adjacent public lands as well as how to be *firewise* in the natural environment.

Mitigation Saves House From Fire

The following is a real-life story about the actions taken by one couple and the end result following a devastating wildfire.[5]

When John and Cindia Hogan bought their home in Los Alamos, New Mexico, in 1994, they did so with knowledge that there might be a major fire in the Santa Fe National Forest that backs up to their property. Their two-story, wood-frame 2,600-square-foot home is located on two-thirds of an acre on Arizona Street.

---

[5] From the Federal Emergency Management Agency (FEMA) website "Mitigation Success Stories," at: http://www.app1.fema.gov/mit/mitss.htm.

John Hogan, a physical scientist with the U. S. Geological Survey, and a trained, experienced firefighter, began taking steps to mitigate their home in 1996. The most valuable information that Hogan had was his knowledge of landscape ecology, which is based on his position and work with the USGS. He works with vegetation studies and fire history as well as changes in landscape.

John Hogan contracted to have a metal roof put on his home. He also cleared some 100 trees from the rear portion of his property, and has removed flammable materials from his backyard. The cost of mitigation is estimated at about $50,000. All costs were borne by the Hogan family.

On May 10, 2000, the Hogan family evacuated from their home, and on May 11 the Cerro Grande fire—largest wildfire in New Mexico history to date—burned through their neighborhood and other areas of the community of Los Alamos. For two days, the Hogans believed their home was also consumed by the blaze, which burned and destroyed well over 200 homes, leaving more than 400 families and individuals homeless.

When John and Cindia Hogan returned to their home, they found it and one other adjacent house intact. Homes to the west and south of them had been destroyed. "We're very conscious of fire danger," Hogan said. "We consciously chose fire mitigation as a proper move."

The Cerro Grande fire caused one tree in the Hogans' front yard to catch fire, and burned a shed and its contents at the far rear of his backyard. The only other damage from the fire was soot that entered into the dryer vent.

The home is insured for $270,000 for its structure and another $200,000 for contents. The savings, even though his property is insured, is figured at more than $450,000—the value of the structure and its contents—or nine times the cost of mitigating the structure and grounds.

Hogan plans to do more mitigation, including removal of more trees in his yard, and put fire retardant on cedar shake paneling on the east and west walls of his home.

"It was certainly worth it," Hogan said of the expense of mitigating his home.

BASIC PREVENTIVE STEPS TO TAKE OUTSIDE YOUR HOME

The key to preventing a wildfire from damaging your property or home is to keep it as far away as possible, and make your structures as defensible as possible. If you already live in the wildland/urban interface, or are thinking of purchasing or building a home there, this section provides practical information on how to protect the outside of your home from the threat of wildfires. Another excellent source of information is the Firewise.org website. Check out their online video "Making Your Home Firewise" along with the accompanying materials on "Firewise Landscaping and Maintenance."[6]

What is Defensible Space?

Defensible space is the "cleaned up" area between a structure (such as your home) and the natural vegetative fuels that exist in the wildland/urban interface. Creating a defensible space involves modifying or removing these fuels and other combustible materials within that space in order to reduce or eliminate a wildfire's ability to spread to your structures. If you have a defensible space around your home, it is less likely that an approaching wildfire will be able to reach your home because it will run out of fuel when it reaches the defensible space. By removing obstacles from the defensible space, you also make it easier for someone—you, your family, or firefighters—to defend your home from a wildfire.

However, some people mistakenly believe the term "defensible space" means that firefighters will automatically show up to *defend* their homes from an approaching wildfire. This is not the case. First, having an adequate, well-planned, and properly maintained defensible space is a form of *self defense,* making your home less likely to *need* defending by people. Second, during a wildfire emergency, firefighting resources are stretched to the limit and it becomes impossible to send a fire truck and crew to every home that is threatened. Third, having a defensible space around your home that is free from obstacles and potential traps may allow firefighters who are available in your area during a wildfire emergency to decide they are able to defend your home safely and thus more likely to do so.

This confusion over the term "defensible space" has led some fire authorities to look for other terms to describe this concept. You may

---

[6] See http://www.firewise.org/pubs/fwc/mainprez_index.html. See also the interesting and informative "Firewise Landscaping" site at http://www.firewise.org/pubs/fwl/contents.html.

therefore see it referred to as a "hazard mitigation zone," a "home ignition zone," or a "survivable space." Since "defensible space" is still the most common term in current use, and for the sake of clarity and brevity, we'll use that term for our discussion purposes here.

General Considerations in Creating a Defensible Space

"Wildfire will find the weakest links in the defense measures you have taken on your property. The primary determinants of a home's ability to survive wildfire are its roofing material and the quality of the 'defensible space' surrounding it. Even small steps to protect your home and property will make them more able to withstand fire."[7]

The better you plan and implement defensible space, the better the chances are that your home will survive a wildfire. A well-designed defensible space will also give firefighters room to maneuver if they need to battle a blaze on your property.

According to a recent study by the College of Agriculture and Life Sciences at the University of Arizona:[8]

"Creating defensible space around your home is one of the most important and effective steps you can take to protect yourself, your family, and your home from catastrophic wildfire.

"All vegetation, naturally occurring and otherwise, is potential fuel for fire. Its type, amount and arrangement can have dramatic effects on fire behavior. There are no "fireproof" plant species. Plant choice, spacing and maintenance are critical; where and how you plant can be more important than what species you use. However, given options, choose plant species for your landscape that are more fire resistant."

In general, defensible space is an area around your home that, once cleared of physical obstructions and potential fuels, provides a "fuel break"

---

[7] Copyright 1999 Colorado State University Cooperative Extension. Reprinted with permission from Colorado State University Cooperative Extension, from 6.302, "Creating Wildfire-Defensible Zones" by Frank Dennis, Wildfire Hazard Mitigation Coordinator, Colorado State Forest Service, http://www.ext.colostate.edu/pubs/natres/06302.html.

[8] From "Firewise Plant Materials for 3,000 ft. and Higher Elevations" (August 2002), by Tom DeGomez, Jeff Schalau, and Chris Jones, http://www.ag.arizona.edu/pubs/natresources/az1289.pdf.

to keep the fire as far away as possible for as long as possible. You can create defensible space by removing or minimizing untended grasses, weeds, brush, trees, trash, scrap wood, firewood, deadfall, abandoned structures, and other potential fuels to lessen a fire's intensity and keep it away from your home. The more of this undesirable material you clear away, the greater the chances that your home will survive.

> "But though the flames jumped across a firebreak Swanson had cut into his backyard, they fizzled out where he had hired crews to thin and prune trees and brush. This 'defensible space' slowed the fire and gave firefighters room to work."[9]

Defensible space does not mean space that is barren; rather, it is space around structures that has been landscaped, generally with native species, but also with other fire-resistant plants, trees, shrubs, and green, well-watered and cut grasses, as well as cleared open space. It is an area that is free of debris, deadfall, and trash. It is also free of outbuildings (other than perhaps a garage).

> "Many people resist creating defensible space around their homes because they believe these areas will be unattractive and unnatural. This is far from true. With careful planning, *firewise* landscaping can be aesthetically pleasing while reducing potential wildfire fuel. It can actually enhance beauty and property values, as well as personal safety."[10]

The following diagram illustrates how the proper use of fire-resistant plants and other materials in landscaping creates a "defensible space" around the home. The key elements in developing fire-resistant landscaping are:

- Adequate defensible space of at least 35 feet from your house (varies and may be up to 100-150 feet depending on the slope of the property).
- Use of fire-resistant vegetation and non-combustible landscaping materials.

---

[9] From "The Other Firefighters" by Bryan Foster, High Country News, Vol. 34, No. 16 (September 2, 2002).
[10] Copyright 1999 Colorado State University Cooperative Extension. Reprinted with permission from Colorado State University Cooperative Extension, from 6.305, "Firewise Plant Materials" by Frank Dennis, Wildfire Hazard Mitigation Coordinator, Colorado State Forest Service, http://www.ext.colostate.edu/pubs/natres/06305.html.

Defensible Space and Fire-Resistant

1. Dispose of stove or fireplace ashes and charcoal briquets only after soaking them in a metal pail of water for 24 hours.
2. Store gasoline in an approved safety can away from occupied buildings.
3. LPG tanks should be far enough away from buildings for valves to be shut off in case of fire. Keep area clear of flammable vegetation.
4. All combustibles such as firewood, picnic tables, boats, etc. should be kept away from structures.
5. Garden hose should be connected to outlet.
6. Clean roof surfaces and gutters regularly to avoid accumulation of flammable materials.
7. Remove portions of any tree extending within 10 feet of the flue opening of any stove or chimney.
8. Maintain a screen constructed of non-flammable material over the flue opening of every chimney or stovepipe. Mesh openings of the screen should not exceed 1/2 inch.

Landscaping Illustration[11]

9. Shrubs should be spaced at least 15 feet apart.
10. Remove branches from trees to a height of 15 feet.
11. A fuel break should be maintained around all structures.
12. Have fire tools handy such as: ladder long enough to reach the roof, shovel, rake, and bucket for water.
13. Each home should have at least 2 different entrance and exit routes.
14. Names of roads should be indicated at all intersections.
15. Names and addresses of occupants should be posted at driveway entrance.
16. All roads and driveways should be at least 16 feet in width.

---

[11] Illustration provided by the National Interagency Fire Center (NIFC) and the Advertising Council. Used by permission.

The information in this chapter and this illustration are representative of the defensible space/fuel break zone requirements of several wildfire-prone states. The most suitable approach for your area will likely be somewhat or very different, but the essential principles are the same.

Finding specific, detailed information about what constitutes "defensible space" can be quite a challenge. Some states have clearly stated requirements, but even those requirements can be further specified by your municipality, county, federal agencies, and many other fire authorities and regulatory entities. Some states have only minimal written policies. In many cases, policies that exist on the books are poorly communicated to the affected parties or not enforced. Be proactive by doing your homework and checking with your local planning commission, building department, or county extension office.

The best way to find out the requirements for compliance where you live is usually to talk to your state, county, and city fire authorities. They are very good sources of information and many will have someone available to walk your property with you and point out what you can do to maximize your chances of surviving a wildfire.

Author's Note:

When I built my house, I had to "earn" a certain number of "points" to get a certificate of occupancy, and one of the main contributors to the point total was clearing trees and deadfall at least 35 feet from any structure. Since I live in the foothills near Denver in a heavily wooded area at the edge of the wildland/urban interface, I asked my local fire department to inspect my property to verify that I had adequate defensible space. I was told that my defensible space was adequate, particularly in terms of appropriate tree distance from my house, but that I needed to clear away some deadfall. The inspector mentioned that if there was a fire, homes with no defensible space—trees grown right against them, firewood stacked against or near them, lots of deadfall everywhere—would likely have to be let go. Firefighters would be safer around my house and more likely to be able to save it.

If you live in or near Bureau of Land Management, USDA Forest Service, or similar lands and do not have adequate defensible space, firefighters might make little or no effort to save your structures. If you are unsure whether this applies to you, visit the appropriate state and federal websites listed in the Webliography and e-mail your concerns to the "contact us" option.

Designing Your Defensible Space

As soon as you are clear about what fire code requirements must be met, you can begin planning how you will comply. To help you decide what is best for your area, consider the following:

- Use the services of a landscape specialist to design your plan, and then get the plan approved by your local fire marshal or similar authority.
- Check with your state, city, county, federal, and other government resources (found in the blue pages of your phone book or via online search) to find out what plants, trees, grasses, and shrubs are best suited to your area.
- Visit your local plant nurseries, county extension demonstration gardens, botanic gardens, arboretums, and similar places to see what grows well where you live and to collect landscaping ideas. Many of these places have "master gardeners" who can help you finalize your plans for free.

Although every area of the country will vary, there are a few general things that you may want to consider in your defensible space:

- If you have sheds or other outbuildings, they should be built or covered with flame-resistant or noncombustible materials, including their roofs. They should also be far enough from your main structures that they don't impede the ability of the firefighters to reach your home.
- Decorative landscaping materials such as rock, gravel, slate, or other non-toxic and non-flammable materials are great ways to create visual interest while breaking up the areas of vegetation and providing access paths to, and escape routes from, your structures (something firefighters really appreciate). These materials can also act as fire barriers, slowing or preventing fire from reaching your structures.
- In general, make sure the plants closest to your structures are widely spaced and not very tall (under 12 inches is best). Vegetation that is farther from your structures can gradually increase in height, density, and variety. It's best not to plant anything within three feet of a structure.
- Plant in selective clusters, rather than in dense groups.
- Keep shrubs outside your defensible space.
- Plant grasses only if they will be kept green and well mowed.
- Non-flammable patio containers are a great way to add color and conserve water.
- Go for variety and maximum use of native species. Plants that are supposed to grow in an area are more likely to resist disease and insects,

and so be more likely to remain healthy. Native species usually come back each year and they almost always recover faster from a fire than non-native species.

A former Assistant Fire Chief in the Oakland Fire Department pointed out the danger of one of the most prevalent trees in the area—the many types of eucalyptus tree. These trees, native to Australia, have bark that peels (much like the sycamore trees found in many parts of the United States). During the five year drought that preceded the 1991 Oakland Hills fire, the eucalyptus were so dry that, to paraphrase this former Chief, the trees went up in flames as if they had been soaked in gasoline. Residents reported hearing loud explosions—the trees exploding and transformers disintegrating.

Using Fire-Resistant Plants and Landscaping Materials

When designing your defensible space, fire-resistant vegetation and noncombustible materials make the most sense. A combination of the two can create a buffer that both helps protect your home and creates an attractive setting.

Fire-Resistant Plants

Just what is considered a fire-resistant plant? According to a recent study by the Oregon State University Extension Service, fire-resistant plants are those that don't readily ignite from a flame or other ignition sources. Although fire-resistant plants can be damaged or even killed by fire, their foliage and stems don't contribute significantly to the fuel load and, therefore, the fire's intensity.[12]

Plants that are fire-resistant have the following characteristics:

- Leaves are moist and supple.
- Plants that have little dead wood and tend not to accumulate dry, dead material within the plant.
- Sap is water-like and does not have a strong odor.

Non-combustible Landscaping Materials

Non-combustible materials such as gravel, rock, paving stones, brick, and similar items are perfect ways to separate structures from ground cover, create walkways, and divide the terrain.

---

[12] From "Fire-Resistant Plants for Oregon Home Landscapes" (April 2002) by Stephen Fitzgerald and Amy Jo Waldo, http://extension.oregonstate.edu/deschutes/FireResPlants02.pdf.

Flagstone, brick, concrete, and even "dyed" concrete, combined with selected ground cover like succulents, can create attractive rock gardens that help protect your home while adding to the value of your property. Fishponds, outdoor sculptures, stone benches and metal patio furniture create an eye-appealing defensible space.

If you are building a deck that is above ground level, one option is to use synthetic materials rather than pressure-treated lumber, redwood, or other traditional materials. A deck made from synthetic materials (plastic-wood composites) like Trex® or "vinyl lumber," will last longer than most wooden decks and require far less maintenance. And while still somewhat combustible, synthetic decking materials probably won't burn as fast as a weathered wooden deck. Be aware, however, that fire researchers have given mixed reviews to synthetic decks. While they do not ignite readily, they can melt (causing issues for firefighters who might fall through them). They also hold sustained "glowing combustion" when covered by embers for an extended length of time, which may cause the adjacent wall of the house to ignite. Accordingly, Firewise.org no longer advocates using synthetic materials for deck construction.

The ideal solution, if your deck is at or near ground level, is to build or rebuild it with brick, flagstone, or concrete—materials that do not burn.

Fire resistant walkways around your home and other structures make escaping from the area easier and safer, and they provide a barrier to encroaching fires. They are also helpful to firefighters since a well-laid stone or rock pathway is far easier to walk on than loose scree and soil.

If you have steep slopes on your property, consider building retaining walls that "stair-step" down the slope to reduce the effective steepness. Steep slopes make it easier for fire to reach the next tree crown and climb the "ladders" created by shrubs. Retaining walls made of stone or other noncombustible materials also provide a fire barrier. It is best to avoid building walls out of tree trunks or untreated railroad ties. Treated wooden ties may be less likely to burn, but there is a danger that the fire may smolder.

Choosing the Right Plants

Whether you are adding plants or need to eliminate unwanted plants, this section will help you determine those that minimize your risk of being overrun by a wildfire. There are five main types of plants to consider for your defensible space and what should be near your home and other structures; ground cover, grasses, flowering plants, shrubs, and trees.

Ground Cover

When selecting ground cover, your best choices should be native plants (vegetation), as they are more likely to meet the criteria for fire-resistance detailed above. Most ground cover plants are easy to grow and maintain. Many are succulents (such as cacti) and require little water, even during a drought. When combined with rock, gravel, stepping-stones, or other noncombustible materials, they can be used to create an attractive alternative to highly flammable grasses, weeds and wildflowers and provide a very effective barrier to the spread of fire.

Plants suitable as ground cover are low growing; generally no more than 12 inches in height. As these plants spread across the ground, their roots and foliage become increasingly dense, reducing the likelihood of soil erosion, especially on steep slopes. In addition, ground covers generally help keep out weeds and other undesirable plants.

Another form of ground cover you might consider including is mulch. Try combining organic and inorganic mulch by placing compost or similar material directly on the soil and around the plants and then covering this with inorganic ground cover such as crushed rock or gravel. Organic material has a tendency to smolder if exposed to fire, so you'll need to be careful not to use too much. If you live in an area frequented by bears, you may want to buy, rather than make, the compost, as bears are attracted to many of the "ingredients." Make sure the mulch is not something that is combustible itself; that means no bark, pine needles, dead leaves, deadfall, or similar materials.

Grasses

Grasses are easy to ignite unless they are very green, moist, and well watered, so they should be kept well away—at least three feet—from any structures. In most of the country, wildfires can occur almost any time of the year, and dry grass is a quick way to bring fire right to your front door and straight up your trees and shrubs. Grass is not a good choice between paving stones or to separate flowerbeds, as it creates a "bridge" in your otherwise non-flammable materials. If you must have grass near a structure, keep it short, green, and well mowed.

Flowers

Wildflowers and domestic garden flowers are lovely to look at, but they also provide potential fuel for wildfires. When they die back in late summer, all the way through to late spring, they are simply another form of readily available fuel to help spread a fire. Clear out the deadfall regularly so that it does not accumulate. It's also best to plant flowers well away—at least three feet—from any structures, allow for lots of space between plants and to have non-combustible materials between flowerbeds. One way to accomplish this is to separate flowerbeds with rock walls, paving stones, crushed rock, or gravel. A better way to add color is to use potted plants, making sure to keep them pruned and watered. Spring flowering bulbs are an excellent way to add color. They require little water, die back, and can be trimmed before the fire season usually gets started.

Shrubs

Shrubs, such as juniper, Gambel (scrub) oak, mountain mahogany, chokecherry, and raspberry, provide food and shelter for wildlife, but they are also a key factor in the spread of wildfires. They are often referred to as "ladder fuels" because they are just tall enough to give ground fire a "lift" into trees. Once a fire spreads into the treetops (a "crowning" fire), it is very difficult or impossible to control. Shrubs can be highly flammable and therefore very dangerous components of the "fuel continuum." Their presence greatly increases the possibility of wildfires spreading and, if close to a structure, can "lift" a fire right up to the roof.

In addition, shrubs are frequent sources of "firebrands"—bits of flaming material that are carried in the smoke column that precedes a fire. These firebrands spread a fire through "spotting"—"seeding" new areas of fuel ahead of the main body of fire. This means that shrubs can eject firebrands that "seed" fire in the grass, trees, trash, and/or deadfall around homes and other structures, as well as on the structure itself.

According to Frank Dennis, Wildfire Hazard Mitigation Coordinator, Colorado State Forest Service:

> "To reduce the fire-spreading potential of shrubs, plant only widely separated, low-growing, non-resinous varieties close to structures. Do not plant them directly beneath windows or vents or where they might spread under wooden decks. Do not plant shrubs under tree crowns or use them to screen propane tanks, firewood piles or other flammable materials. Plant shrubs individually, as specimens, or in small clumps apart from each other and away from any trees within the defensible space. Mow grasses low around shrubs. Prune dead stems from shrubs [at least] annually. Remove the lower branches and suckers from species such as Gambel oak to raise the canopy away from possible surface fires."[13]

---

[13] Copyright 1999 Colorado State University Cooperative Extension. Reprinted with permission; from 6.303, "Fire-Resistant Landscaping" by Frank Dennis, Wildfire Hazard Mitigation Coordinator, Colorado State Forest Service, http://www.ext.colostate.edu/pubs/natres/06303.html.

Shrubs can add greatly to the beauty and value of your property, but if you live in an area prone to wildfires, you need to choose and plant them with great care.

Trees

Trees, especially evergreens, are an excellent source of fuel for wildfires. They often eject firebrands (as do shrubs) that are carried well ahead of a fire, causing spotting. In addition, trees that are already burning emit such large amounts of radiant heat that nearby grasses, plants, shrubs, trash, and other trees, as well as buildings, may appear to undergo spontaneous combustion.

When choosing trees to plant on your property, or deciding which trees to remove and which to keep, plan on having *no* trees closer than 35 to 100 feet from any structure (check your local ordinances for specifics). As a rule of thumb, you need to know the full adult height of any trees you plan to use and plant them at least that distance from any structures. If your site is already treed, consider this rule when locating and building new structures. Call your local tree maintenance service for assistance, or visit your local arboretum.

Author's Note:

"The biggest concern I hear property owners voice is that their privacy will be taken away if the trees are cleared back to a safe distance. I don't agree—I cleared mine to 35 feet or more and still have one of the most secluded houses in the area. My trees are also less dense and healthier, and seem to have weathered the drought better than others in the area."

Take a look around your property to see what grows best when choosing trees. If deciduous trees such as oak, maple, aspen, larch, and so forth will grow in your area, consider planting them rather than coniferous trees. Deciduous trees (those that lose their leaves every fall) tend to be more fire-resistant than coniferous (evergreen) trees and are an excellent *firewise* landscaping choice. They can actually provide some shielding from a wildfire's intensity.

If you must use evergreens, make sure to plant them so that when they mature (not merely the young saplings) they will have clear separation of several yards between the edges of their crowns. If you plant seedlings and saplings close together, you will need to thin them as they grow. Many evergreens can reach 100 feet or higher and spread to cover a large area, so that means *significant* spacing on the ground. If you are planting evergreens on a slope, plan on at least 20 to 25 feet between the edges of mature crowns.

Maintaining Your Defensible Space

"Our property has been a designated Stewardship Forest for over a decade. Under the management plan we developed with the Colorado State Forest Service, we do prescribed thinning annually, removing diseased trees, and burning slash piles when the weather is favorable. Last year we worked more on defensible space around our buildings including raking pine needles and cutting low shrubs and branches. During the Hayman Fire a fire crew came and spent part of a day doing fire preparation work that included moving all the wood from our woodshed about 100 yards toward the creek, cutting several more trees close to our main cabin and cutting a fire break around the perimeter of the buildings clustered on our property."

— *Candace Boyle*

A well-maintained landscape is essential to creating and keeping defensible space. If you "spring clean" your home and change the batteries in your smoke detectors when you "spring forward" and "fall back," it is also an excellent time to clean up the accumulated deadfall and other stuff that has built up around your property. Plants that are not properly maintained, pruned, and thinned become fuel for fires when allowed to die or become too dense. What can you do to ensure your fire-resistant landscape fulfills its purpose?

- Make every effort to keep plants in top condition, removing the deadfall and pruning to encourage more vigorous growth and deeper root systems.

- As evergreens mature, prune branches to at least 10 feet above the ground (Be sure to consult a tree service, arboretum, or similar entity for advice or assistance and be careful to prune no more than one-third of the tree at a time).

- Dispose of pruned branches. Do *not* stack them by the house to burn later. Tree services normally run branches and trees through a chipper and spread the result over a wide area to reduce fire risk. Proper disposal of pruned and dead vegetation will greatly reduce the "ladder" effect associated with nearby shrubs and grasses.

- Plants have different growth patterns at different times of the year. Remove annuals and perennials after they have gone to seed or when the stems become overly dry. Remove dead stems from bulbs.

- Remove stalks and residue from vegetable gardens as they are harvested or die back.

- Rake leaves and other litter as it builds up through the season.

- Regularly remove dead leaves, pine needles, and other debris from gutters and drains.

- Keep chimneys and flues clean and free of resin and soot build-ups.

- Mow or trim grasses to a low height within your defensible space. This is particularly important as grasses dry out.

- Remove plant parts damaged by snow, wind, frost, disease, and insects.

- Remove or kill weeds adjacent to your structures. They grow quickly and die back, creating fuel that gives wildfire a free ride to your house and a lift up into bushes and trees (the "ladder" effect). It is wise to maintain a three-foot "fire free zone" around your house and any outbuildings. The existence of debris or vegetation growing against the house has brought grief to many homeowners who become victims of wildfires.

- Use and then store gas grills away from your house.

- Do not use the space under your deck to store anything flammable (or things with flammable contents) such as firewood or lumber, lawn mowers, propane tanks, gas cans and paint cans.

- Keep your driveway clear, especially if it is a long one that runs through trees. It can act as a firebreak.

- If you live in an area that is experiencing drought, you may have to give up most of your vegetation. Native plants are often drought-tolerant or at least more drought-resistant than imported varieties. But, if you must let your landscaping die because of watering restrictions, remove the deadfall regularly. Remember that mulch is a great way to conserve moisture around your plants, but it needs to be fire-resistant, too. Dead pine boughs, bark, and needles are *not* good for mulching your defensible space—they are highly *flammable*.

Even the best defensible space can be destroyed if you don't keep it up. The old adage "An ounce of prevention is worth a pound of cure" certainly applies here.

CHOOSING THE RIGHT HOME AND PROPERTY

Living in the wildland/urban interface, or actually inside a state or federal wildland area, is the dream of many potential homeowners. Whether you are buying or building, picking the most *firewise* home and property from the beginning could save you thousands of dollars and the work required to do the things listed in this chapter. As you shop for your dream home, here are some things to consider.

Understanding the Rules

To begin, if the property adjoins public land, contact the responsible government agency and ask about its management goals. They will also be a great source of tips and ideas to make your property as fire resistant as possible. If you are not actually inside a USDA Forest Service or BLM property, your local government, fire authority, or state forester probably sets the rules for fire protection compliance, and they would be your best source of information.

If the community you are looking into has a homeowner's association, it is best to contact the association and get copies of their by-laws and covenants. From these documents you will know if there are any requirements or restrictions that would prevent you from following the steps you want to take to make your property *firewise*. See if the covenants allow you to create a defensible space. Sometimes homeowners' associations unintentionally cause or compound the "un-defendable" problem by requiring homes in their areas to have cedar shake roofs or landscaping that contains flammable vegetation and materials. For obvious reasons, such policies and covenants are unwise in the wildland/urban interface. If you have picked a home in one of these areas and you cannot feasibly implement the necessary actions, especially regarding defensible space, you should consider buying somewhere else.

Adequate Water Supplies

Even if the actual property has a well, it is likely not adequate for fire suppression. In my neighborhood, we have two additional water sources: a large container of water placed by the county and strategically placed cisterns. If there is access to a creek, that can be supplemental water for firefighters, but keep in mind that natural water sources may run low during certain times of the year and in extended drought cycles. If the home does not have outside water spigots, consider the cost of a plumber to have them added.

Survivable Escape Routes from the Property, Subdivision, and Fire-Danger Zone

Most fire codes require at least two completely separate entrance and exit routes from both the house and the subdivision. If one of the subdivision exits requires a four-wheel drive vehicle, and you don't own one, you would need to weigh the cost of buying one against the risk of not having one. Many of the people where I live manage just fine all winter with front-wheel drive vehicles, but just recently our secondary exit route required a four-wheel drive to negotiate it. When a local wildfire loomed over the other side of I-70, we were told if we had to evacuate we should plan to use the alternate route. You can imagine the panic among the people with inadequate vehicles. If there is only one way in and out, consider buying elsewhere.

The Land and the Terrain

While a home way up on the highest, most private, least accessible part of the property may be your ideal, the reality is that it may not be accessible to the professionals who fight fires, or it may be too far from a reliable water source.

If the property is in an area prone to fire, floods, or mudflows, you'll want buildings to be constructed and positioned to minimize those risks. Check the slope of the property around the house. If it is more than 6 percent, you may want to consult a geotechnical engineer for a slope-stability study.

Make sure you are not on or near an abandoned (or working) coal mine. These mines have been known to catch fire and smolder for years and then erupt, destroying homes and property—as occurred with the Coal Seam Fire near Glenwood Springs, Colorado in 2002.

Fire Resistant Materials

If the house you might buy has cedar siding or a wooden deck, make sure it has been treated with a fire retardant chemical within the last year, or try to add that to your closing costs. Keep in mind, however, that fire retardants have a limited lifespan and need to be reapplied periodically according to the manufacturer's directions. Further, fire retardants only reduce the flammability of wood and other combustibles; they do not eliminate it. Never assume that treated wood is *fireproof.*

If the roof of the house is made of wooden shingles or other combustible materials, your insurance company may not insure the house. Since most lenders require insurance as a condition of the loan, you may need to replace the roof with a nonflammable material like tile or metal before you can sign for the loan. Composite roofing materials, such as asphalt shingles, may be acceptable if they have a good fire rating. Firewise.org recommends a "Class A" rating. A reputable roofing contractor can provide you with the necessary information to make this determination.

SPECIAL CONSIDERATIONS FOR NEW HOME CONSTRUCTION OR REMODEL

If you are building your dream house in the wildland/urban interface, or you are planning to up-date or repair your existing property, there are many things you can do to make your home as fire-resistant as possible. Most of these actions do *not* add dramatically to the cost of the structure. Some may even decrease your insurance rates and improve the energy efficiency of your home. In addition, you will want to implement as many of the actions discussed in the "Basic Steps to Take Inside Your Home" section as you can afford. Again, check out the "Making Your Home Firewise" video on the Firewise.org website, along with the accompanying materials on "Firewise Location" and the "Features of a Firewise Structure."[14]

First and foremost, it is important *always* to comply with your state, local, county, city, and other building codes. If you do not cooperate with the building inspectors and pull the right permits, you may never get a certificate of occupancy and your insurance may not cover losses due to non-compliance. Make sure you get your local fire authorities involved from the beginning and follow all their requirements. Most authorities truly want to help and are just waiting for you to ask.

Many jurisdictions have strict building codes and rules for getting a certificate of occupancy or for passing inspections in a remodel. In some places these codes and rules are clearly stated, and in others they are not. If you are building in a county where the fire codes are fairly loose, check the codes and rules in an adjacent, stricter county with similar wildfire risks and use them as *your* criteria instead. If you have to meet a "point" system for fire safety, there are usually multiple ways to satisfy the point requirements, especially for smaller houses. Generally speaking, the rules are fairly easy to meet for homes under 2500 square feet (total including

---

[14] See http://www.firewise.org/pubs/fwc/mainprez_index.html.

basement), a bit harder for homes of 2501-3500 square feet, and increasingly hard to satisfy with larger homes. Once you reach 5,000 square feet or more, you may be required to have a fire-resistant roof, 5/8 inch drywall, a fire-suppression sprinkler system, monitored alert systems, and other safety devices.

Things You Might Do Outside

- Site your house in the most favorable position. Walk the site with your local fire department or state forestry office representative and include your builder and architect on this tour of inspection.

- Situate and landscape the house with defensible space in mind.

- Plan for adequate drainage for runoff to prevent floods, mudflows, and landslides now, or if a wildfire occurs.

- Consider installing an automatic underground sprinkler system for your yard. Besides making ordinary watering easier, a sprinkler system will help you keep your defensible space green and moist, and therefore less susceptible to fire.

- Develop a plan for thinning trees and removing debris and deadfall. This is relatively simple to do before or during construction, but it may be difficult or impossible once your home is built.

- To provide access to your home to fire and other emergency vehicles, make sure your driveway meets or exceeds width, weight, and any other code requirements. If feasible, make your driveway a loop with two different exits onto the road.

- Consider having a detached garage—it may survive even if the other structures do not.

- Put your utility lines underground. Underground lines require little or no maintenance and are less likely to be damaged by fire, weather, or people. Damaged gas or electric lines may actually cause a fire. The same is true for propane tanks, which may also be installed underground.

- Make sure your address is clearly visible from the road. Consider having address placards made out of nonflammable material such as stone or metal. Place them at the beginning of your driveway and along the road directly in front of or across from your main building or home.

Things You Might Do to the Structure

To make your house *firewise,* you may have to make some trade-offs. You may have had your heart set on a cedar shake roof, but the fire department or insurance company has said no. Fortunately, there are lots of alternatives to choose from. A good place to collect ideas and cost information is at "home and garden" or "home building" trade shows found in most metropolitan areas. Here are some of the things you might want to do to make your home safer.

- Consider tempered safety glass for larger windows and patio or sliding doors. Your building code may even require this. Glass block may be another option—and it now comes in pre-fabricated Plexiglas forms so it is uniform and easy to install.

- Put in a propane or gas burning stove rather than a wood burning or pellet stove or fireplace. The cost is minimal and the energy efficiency is a great cost savings. My two propane stoves can heat the entire main level of my house, and raise the temperature 20 degrees Fahrenheit in less than 30 minutes.

- Make sure there are several outside water spigots.

- Make sure there are at least two exits on the main level and one or more on the lower (basement or walkout) level.

The Safest Roofing

Choosing the right material for your roof could be the single most important decision you make. Embers blown onto your roof can be the start of a blaze, even when you have taken the steps for defensible space. There are many attractive, nonflammable roofing materials available now—tile, metal, cedar shake look-alikes—that have the same look without the fire risk. Check with fire officials and manufacturers for fire ratings to help you make a choice. Firewise recommends Class-A asphalt shingles, metal, cement and concrete products, or terra-cotta tiles.[15]

Author's Note:

"I have a metal roof and love it. It should last 40 years or more; it is bolted down and secure from the wind; it is almost impervious to hail; the color does not fade, and it greatly adds to my home's fire resistance. And no, I don't hear the "pattering of rain on the roof" even though I have cathedral ceilings. A metal roof probably will

---

[15] See http://www.firewise.org/pubs/fwc/roofing.htm.

cost twice as much as a composite roof but less than a tile or other noncombustible roof; but costs vary widely across the country."

Exterior Siding

Where you live, average humidity, fire codes, and even your community association may determine what outside materials you choose for your home. Vinyl siding is not a good choice, in any event, because it melts even from radiant heat. Local covenants may require you to use wooden siding. If you choose or must use wood products on the outside of your house, be sure to apply fire-resistant stains or treatments to the exterior surface and maintain the siding as the manufacturer recommends. Cedar shake siding (as opposed to roofs, which are not recommended) can actually be somewhat fire resistant—if the shakes are in good condition and are treated regularly (every year) with fire-resistant chemicals. On the other hand, dry faded wood is like tinder if it is not maintained and treated regularly. It is important, too, not to let pine needles and debris build up at the base of exterior walls or in the gutters above them. Other siding choices like metal, brick veneer, stucco, and stone materials are fire resistant and do a better job of protecting your home. You may find that any added cost for these types of materials is repaid, in part, through lower insurance rates, having a home that firefighters can defend, and one that is still standing after a fire.

Consider Thicker Drywall

Most construction uses 1/2-inch drywall. Increasing this to 5/8-inch dramatically increases the ability of your building to resist fire (to almost double). If you are on a point system, this is often worth as many points as a sprinkler system, at a much lower additional cost. If you add this to 2x6-inch rather than 2x4-inch framing, you should see several benefits: greater burn resistance, better sound isolation between levels, and lower heating and cooling bills due to better insulation. Depending on where you live, the cost of installing that extra 1/8-inch of drywall may be double that of 1/2-inch drywall.

BASIC STEPS TO TAKE INSIDE YOUR HOME

There are some internal modifications you could make to your new or existing home or other structures that are very basic, but often overlooked. Detection and alarm systems (if properly placed) are things that will alert you if a problem has begun whether the fire has started inside or outside. In new construction or a renovation, building codes may require some or all of these devices. Once they are installed, it is important to keep them monitored and in good working order, so regular testing is key to making sure they are there when you need them. The main systems to consider are listed below.

Smoke Detectors

Smoke detectors can be stand-alone battery-operated detectors or ones that are hard-wired into your building's electrical system and other alarm systems, with a battery back up. Hard-wired detectors are especially convenient since you don't have to rely on batteries or test them regularly to ensure they are working. Detectors should be on every level of your home, including the basement and attic, and adjacent to all sleeping areas (including areas only used occasionally for sleeping, such as the living room). Do *not* put them in the kitchen or by any location that emits lots of steam (such as inside or right outside a bathroom or near your furnace). Use heat detectors in these areas.

Heat Detectors

Heat detectors are used in the kitchen, where smoke detectors are too likely to be set off by burnt toast. They are used in or outside areas such as bathrooms, where steam might be mistaken as smoke by smoke detectors. Heat detectors can also be installed as stand-alone battery-operated detectors or hard-wired into your electrical system.

Carbon Monoxide Detectors

These detectors should be installed and integrated in the same way as smoke detectors. They normally go outside of your furnace room, but you should ask the fire authorities where else to put them.

Moisture Detectors

While not a major aspect of fire avoidance, these detectors are valuable for preventing or reducing water damage that may result during or after a wildfire. Consult the installation company about the best locations for these monitors.

Make Sure All Your Detection Systems are Remotely Monitored

Monitored detection systems are great favorites with insurance companies and fire departments. They also add greatly to any required point totals. You may save almost enough on your homeowner's insurance annual premiums to pay for the monitoring service. Be aware, however, that monitoring services may not be available in very remote areas.

Consider a Sprinkler System

A fire-suppression sprinkler system inside your home will earn lots of fire-safety points, but your area may not have sufficient water flow or pressure to support a sprinkler system, especially at a time when firefighting efforts nearby may already be taxing the water supply. Nonetheless, if you build a large home (generally over 3,500-5,000 total square feet), you may be required to install one.

Spark Arrester in the Fireplace

If there is a wood-burning or pellet stove in the house, a spark arrester on the chimney can be an excellent, low-cost improvement. If you are in the process of buying a home that does not have one, you may be able to get the current owner to do this, particularly if the building code requires it.

Fire Extinguishers

Keeping an A/B/C-labeled fire extinguisher in rooms that are most prone to catching fire is always a good idea. The kitchen, garage, craft or work shop areas are the obvious ones, but if you have out buildings where flammable materials are stored or used, they are also good choices.

WILDFIRE PREVENTION WHILE ENJOYING THE GREAT OUTDOORS

Fire can be devastating when it gets out of control, but it can be a pleasant, friendly source of warmth, cooking, and other activities when used wisely. If you are one of the millions of people who plan to spend time camping, hiking, biking, backpacking, or otherwise enjoying the many parks, open spaces, wilderness areas, and forests in our beautiful country, there are many things you can do to avoid causing a wildfire.

The best way to minimize wildland fires is to keep them from starting in the first place. Outdoor fire safety is key to preventing wildfires. Of course, if you *see* a wildfire, immediately call 911 and notify the closest federal, state, or local authority you can find.

It is interesting to note, according to the Washington State Department of Natural Resources (DNR), that about 85 percent of all wildfires occurring on DNR-managed land in Washington, from 1958 to 1995, were started by humans. Moreover, these fires tended to occur in the most populated regions of the state. Lightning started only about 13 percent of fires occurring on DNR-managed land.[16] Although lightning ignites a somewhat higher percentage of fires in wilderness areas overall, the point is that people do start fires, and people can prevent them by observing safe burning rules.[17]

Some of the simplest things you can do to reduce the risk of a wildfire from happening in the first place are:

- Always build fires a safe distance from trees, bushes, or other flammable materials and fuels.

- Always have at least one means available to extinguish a fire quickly (whether it is a fire you meant to start or an unintentional one), both indoors and out.

- It may seem obvious, but never leave any type of fire unattended. This means cigarettes, grills, campfires, propane stoves, fires in your fireplace at home, trash and scrap fires, farmland and grassland stubble burning, firecrackers, and so forth. Remember that even a fire that is controlled inside your house can send up sparks through the chimney or onto the floor and eventually make its way into the surrounding wildlands; the house fire then becomes a wildfire.

Another way to prevent wildfires is to exercise extreme caution when you are in fire-prone areas such as forests, grasslands, meadows, and so on. This means using only approved grills and fuels, taking extreme care with any fire source, obeying restrictions (such as bans on open-flame fires and fireworks) posted or publicly announced by local law enforcement, fire, or land management authorities, always making sure all fires are out, and by not smoking in these areas.

Open Fires

An open fire is one where the source of fuel and flames are not contained. A fire pit or charcoal grill is an open fire. If you are grilling or cooking outside using a propane tank and camp stove, that may or may not be considered an open fire because these stoves generally do not emit sparks or embers. *But check with your local fire authority to be sure.* If you

---

[16] See http://www.wa.gov/dnr/htdocs/rp/prevent.htm.
[17] See http://www.wa.gov/dnr/htdocs/rp/safeburn.htm.

are not sure, do *not* start a fire of any kind. These rules may change depending on the season and the current fire-danger level, as well as from state-to-state and even county-to-county. Some wilderness areas and open spaces have even more stringent requirements and may ban open fires altogether, including cigarette smoking. It is your responsibility to know and follow all fire regulations. You can find out by contacting the local forest service or fire authority.

Building and Putting Out a Campfire

Campfires should always be built in a metal or concrete fire pit, or in a pre-laid ring of stones, with a metal grate or cover if possible, and not on a steep slope. There should be no vegetation (or debris or firewood) within a 10-foot radius of the fire, and no tree branches hanging within 20 feet of the fire's highest flames. Start small, adding larger material as the fire catches. Push the larger material into the fire as the inside portions burn. Keep the fire small and well within your control. A fire should not be allowed to grow to more than three feet in diameter. Make sure your match is out cold. You should you be able to hold the burned tip of the match between your bare fingers.

Keep a shovel and at least five gallons of water at hand and make sure the person watching the fire knows where these are. An A/B/C-labeled fire extinguisher is always a good idea, too. It may seem obvious, but never, ever, leave the fire alone and never leave children alone with the fire.

Finally, when you are finished with the fire, douse it with water. Make sure all embers, wood, and coals are soaked. Stir the water into the fire to ensure everything is wet. Move the rocks in the fire ring and douse any hidden embers there, too. Mix soil or sand into the embers and continue stirring until all material is cool enough to touch. If it is too hot to touch, it is not dead enough to leave alone. The fire is out only when the ashes are cold.

An adult should always supervise a campfire. The slightest breeze could carry sparks into trees, clothes, or other combustible materials.

Lanterns, Stoves, and Heaters

To refuel a lantern, stove, or heater, wait until it is completely cool and place it on a level, nonflammable surface to fill it. If you spill any fuel, move the item to a new area before lighting it. Securely close and store containers of flammable liquids in a safe place, well away from heat or flame.

Lanterns and propane stoves should never be lit inside a building, tent, camper, or trailer—light them outside. In fact, it is unwise to bring them inside at all. Even with apparently adequate ventilation, carbon monoxide can accumulate in an enclosed space and be deadly. The better lighting choice is a battery-operated or solar-charged flashlight. Cooking on a propane stove should be done outside. Heaters should be used with extreme caution and only with adequate ventilation and proper venting.

Burning Trash and Debris in Your Community

Most communities forbid burning of trash and debris. Others allow it only at certain times of the day or year, or only for agricultural purposes—and then only with a permit and under certain meteorologically favorable conditions.[18]

If you are allowed to burn trash, please remember always to check the weather. Never burn if it is dry or windy, or predicted to become so.

Additionally, if you are permitted to burn trash, have a shovel, at least five gallons (20 liters) of water (or a connected hose), lots of sand or dirt, and an A/B/C fire extinguisher close at hand. Please remember not to burn under trees or near trees or grasses or other combustible materials. Never leave the fire unattended, and make sure it is completely out before you leave the site. If the burn spot is warm to the touch, the fire may not be out.

It is always a good idea to use a fireproof container such as a 55-gallon drum for burning to contain the fire. It may seem obvious, but make sure there are no aerosol cans, paint or other flammables, or toxic materials (such as plastics) in your burn pile. Some recyclable or disposable materials, such as cans, glass, or plastics, do *not* burn and can create toxic smoke. In addition, they may succeed in creating a smoldering fire that is hot enough to ignite nearby trees and grasses.

There are many alternatives to burning. Many communities have a wood-chipping facility to turn trees and similar debris into mulch. In my neighborhood, we rent a chipper several times each year and let all the residents use it. Grass, leaves, and stubble can be turned into compost. Recycling is a way to dispose of materials such as cardboard, office paper, newspapers, and coated papers.

---

[18] See for example, "Safe Debris Burning in Washington Forests" at http://www.wa.gov/dnr/htdocs/rp/safeburn.htm.

Spark Arresters

Sparks caused by equipment or recreation vehicles are one of the major accidental causes of wildland fires. Therefore, vehicles and other equipment intended for outdoor use are required to have spark arresters. Some examples are lawnmowers, chainsaws, portable generators, and cross-country vehicles (ATVs, snowmobiles, and trail/dirt bikes). If you have a vehicle or equipment you intend to use in, on, or near grass, brush, or a wooded area, it must have a spark arrester. In fact, you may be subject to a fine if you use equipment that does not have a spark arrester.

Your state division of natural resources, USDA Forest Service, and State Forestry office are good sources to find out if the machinery you plan to use needs a spark arrester. To make sure the spark arrester is working properly, contact the equipment dealer.

Charcoal Cooking Fires

Charcoal fires should be treated much the same as a campfire if used out in the open. When you are finished cooking, soak all the coals and stir them to make sure they are completely wet. If you cannot comfortably put your bare hand on the coals, the fire is not out.

Smoking

If you smoke, please be aware of the fire danger caused by cigarettes. In the early days of Smokey Bear, cigarettes were one of the chief targets of the fire awareness campaign. Nothing has changed since then.

During certain times of the year and under certain weather or drought conditions, smoking may be prohibited in wildland areas, or it may be permitted only inside of vehicles. If smoking is permitted out of doors, you should keep lighted matches and cigarettes at least three feet from any dry vegetation or other flammable materials. Grind out your cigarette (or other tobacco product) in the dirt, not the grass, or carry a portable ash tray.

Finally, never toss a lighted match or tobacco product from a window (home, vehicle, camper) or tent, or drop them when hiking, biking, or otherwise enjoying the great outdoors.

Where to Learn More About Wildfire Prevention and the Great Outdoors

Much of the preceding material is taken from the USDA Forest Service pamphlet "Outdoor Fire Safety." You can pick up a copy at your local forest service information center.

HOMES DETERMINED TO BE UN-DEFENDABLE

During the Black Mountain fire in central Colorado, a "small" (400-acre) fire resulted in 1700 homes falling under evacuation orders. However, many of these homeowners refused to leave, assuming firefighters would make every effort to save their homes. *What they didn't know is that firefighters will not put their lives at risk only to save structures, especially those without adequate defensible space and firewise construction.* If residents refuse to follow evacuation orders, firefighters can do little to save them if the fire rushes in. According to Justin Dombrowski with the City of Boulder, Colorado Fire Department, up to 85% of the homes evacuated during the Black Mountain fire had been determined to be un-defendable—meaning firefighters would be unable to safely protect the homes from the flames.

Among themselves, firefighters sometimes refer to un-defendable homes as being "red tagged." This is an *unofficial* slang term for structures with insufficient defensible space, inadequate ingress and egress, or that are designed or constructed in such a manner that it would be unsafe to fight a fire on the property.

Your local government and firefighting authorities are under no obligation to contact you and let you know that you are "red-tagged" and which requirements you have not met. Having a certificate of occupancy is no guarantee that you meet the firefighting authorities' requirements; nor is paying taxes.

It's unrealistic to expect a fire truck to show up at every house during a wildfire emergency. Firefighters must make a choice and try to save lives first and then properties that have the best—and safest—chance at success. Firefighters have a dangerous job, and they are not going to increase their risks by trying to fight a fire in surroundings that compound that risk. They do not have the resources to defend every space and *must* prioritize. **It is *your* responsibility to find out the status of your house, and take appropriate steps to fix any deficiencies.**

GUIDELINES FOR BECOMING A FIREWISE COMMUNITY

"Firewise Communities/USA is a project of the National Wildfire Coordinating Group's Wildland/Urban Interface Working Team and is the newest element of the Firewise program. It provides citizens with the knowledge necessary to maintain an acceptable level of fire readiness, while ensuring firefighters that they can use equipment more efficiently during a wildland fire emergency. The program draws on a community's spirit, its resolve, and its willingness to take responsibility for its ignition potential."[19]

Many communities around the country have adopted the national firewise standards and have been recognized under the Firewise Communities/USA program. Becoming a Firewise Community involves the following steps:[20]

Step 1: Contact Firewise

Becoming recognized as a Firewise Community/USA begins with the community itself. A community representative completes an on-line request for contact by a Firewise representative on the Firewise Communities/USA web site, http://www.firewise.org/usa/.

Step 2: Site Visit

At an agreed-upon time, the local Firewise Communities/USA representative schedules a site visit and assesses the site. The visit is coordinated with local fire officials.

Step 3: Community Representatives

Meanwhile, community representatives create a multi-discipline Firewise board/committee. It should include homeowners and fire professionals. Participation by planners, land managers, urban foresters and/or members of other interest groups is also encouraged. Board members should be informed that development of the WUI [wildland/urban interface] plan will take up to six months.

Step 4: Assessment & Evaluation

Upon completion of the site assessment and evaluation of the community's readiness to withstand a WUI fire, the WUI specialist schedules a meeting with the local Firewise board. The assessment and evaluation are presented for review and acceptance. If the site assessment and evaluation are accepted, the process continues. If they are rejected, it terminates.

---

[19] See http://www.firewise.org/usa/about.htm.
[20] From "How to Become a Firewise Community" at http://www.firewise.org/usa/.

Step 5: Create Plan

The local Firewise board uses the report to create agreed-upon, area-specific solutions to its WUI fire issues. All members of the Firewise board must concur with the final decisions. Their recommendations are presented to, and approved by, the WUI specialist. The specialist may, at that time, work with the community to seek project implementation funds, if those are necessary.

Step 6: Implement Solutions

Local solutions are implemented following a schedule designed by the local Firewise board and WUI specialist. A permanent Firewise task force, committee, commission or department is created that will maintain the program into the future.

Step 7: Apply for Recognition

Firewise Communities/USA recognition status is achieved after the community submits its registration form (available at http://www.firewise.org/usa/). A completed Firewise community plan and Firewise event documentation must also be provided to the local Firewise representative.

Step 8: Renewing Your Recognition Status

Recognition renewal is completed annually by submitting documentation indicating continued community participation to the State Coordinator. This can be accomplished by using on-line forms available at http://www.firewise.org/usa/.

These relatively simple steps leading to recognition under the Firewise Communities/USA program can go a long way toward creating a sense of community cohesiveness and cooperation when wildfires strike. We really *are all in this together.* As an added benefit, recognition may help you to get and keep affordable homeowner's insurance.

Firewise.org also recommends the following "Firewise Practices" for homeowners in Firewise communities:[21]

- Excess vegetation on road shoulders is removed.

- Cedar shake roofs are replaced with a non-flammable, Class A alternative.

---

[21] See http://www.firewise.org/usa/hints.htm.

- Driveways, non-flammable walkways, and other pathways can halt the spread of a wildfire.
- Careful spacing of trees and shrubs lowers wildfire potential.
- Rockeries can interrupt a fire's pathway to a house.
- Fuels are chipped/removed immediately after cutting.
- Wood is piled away from the house.
- A three-foot fire-free area is created on all sides of the house.
- Dead leaves and branches are removed from trees, shrubs, and plants within the "home ignition zone."
- The home ignition zone is free of fallen leaves and needles.
- Indigenous wildflowers and native plants are excellent Firewise choices.
- Green lawns and irrigated areas serve as fire breaks.
- Fuels are thinned at the edge of the home ignition zone.
- Deciduous trees (generally a *firewise* choice) are carefully spaced within the home ignition zone.

In addition to the national guidelines above, you can find information on state and local Firewise chapters, as well as additional state-specific guidelines by following the links at http://www.firewise.org/local.html.

Get Involved

A great way to know what to do and what is planned for your area is to become active in your community. Get involved with the boards and commissions that work with government agencies responsible for fire management and "burn prescriptions." This will help you understand what plans have been made to fight wildfires in and around your new home, who will make the decisions, what you need to do to make your house defensible, and when and where to expect prescribed or controlled burns.

Another helpful resource for this information is your local (town or county) newspapers. With these, you'll get a feel for the area, especially the relationship between residents and government entities.

ADDITIONAL RESOURCES

Much of the information in this book is general. You can get more specific information about your area by visiting your state department of natural resources, department of forestry, or similar website. To find these sites, try the following *Google* searches: (1) [your state] + "department of natural resources" + wildfire; and (2) [your state] + "department of forestry" + wildfire. You might also do the same searches and add your city or county name.

At the back of this book, there is also a list of state foresters for all 50 states to help you find information specific to your state.

Here are some additional sources you might try:

- **Firewise.org.** Some Firewise programs are discussed above. A project of the National Wildland/Urban Interface Fire Program in cooperation with the National Fire Protection Association, this multi-agency program encourages the development of defensible space (by advocating reduction and modification of potential fuels in your "home ignition zone") and the prevention of catastrophic wildfire. Their website includes information, checklists, and other resources for making your home and the space around it *firewise*. Here you'll find lists of "Firewise Practices" and "Firewise Plants" best suited for use in a number of individual states. See http://www.firewise.org/usa/. Many states and municipalities also have Firewise chapters. See http://www.firewise.org/local.html. And many towns and subdivisions are part of the Firewise Communities/USA recognition program. See http://www.firewise.org/usa/communities.htm.

- **Federal agencies,** such as USDA Forest Service, http://www.fs.fed.us/fire/, and the Bureau of Land Management, http://www.fire.blm.gov/, as well as interagency and government sponsored organizations such as the National Interagency Fire Center, http://www.nifc.gov/, and the International Fire Code Institute, http://www.ifci.org/.

- **State forestry office.** There is a contacts list of the state foresters for all 50 states at the back of this book, including their names, addresses, phone numbers, e-mail addresses, and websites. The most current list is also available online from the National Association of State Foresters at http://www.stateforesters.org/SFlist.html. You might also contact your state division of natural resources, college of forestry, or college of agriculture.

- **Local agencies,** such as your county extension office or fire department, as well as private tree services, nurseries, and landscapers.

CONCLUSION

Whether you build from scratch, work with a builder on a "house in progress," or buy an existing home, there are many steps you can take to significantly increase the likelihood your home and other structures will be considered "defensible" by firefighting authorities. Following the steps and suggestions in this chapter should improve your chances that your property will escape a wildfire unscathed or in relatively good condition and possibly save your life.

Chapter 2: Preparation

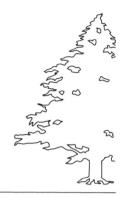

"The fire started on Monday. On Tuesday we were told that the fire was out. Wednesday morning, ash was raining down all around us. My son called and told us that the fire had exploded, we looked and saw it coming over the ridge. We were told to evacuate; we had one hour. Because we had experienced this before, five years ago, we knew exactly what to grab: important papers, some food, clothes, and photo albums. My husband is an invalid, so my daughter and I had to do it all."

— Mrs. Lois Trimble,
resident of Navajo County, Arizona,
on the Rodeo-Chediski Fire (2002).[22]

If you live in the wildland/urban interface, you may already have lived through one or more wildland fire seasons. You may also have seen articles in local newspapers about what you should do and the things you should take with you if a wildfire threatens your area and you need to evacuate. This chapter contains several useful lists and tips for preparing your family and property, including steps you should take long before you face a possible evacuation, what to do if you have little or no time to prepare, tips on how to evacuate in a planned orderly manner, and ways to protect yourself if you are in imminent danger.

BASIC PREPARATION FOR HOMEOWNERS

Long before you face a possible evacuation, it makes sense to do several things:

- **Get your important personal and business papers in order** and put them in a safe place well away from any fire danger, preferably a safe-

---

[22] From Mitigation Success Stories on the FEMA website at
http://www.fema.gov/fima/success.shtm.

deposit box. These papers include original birth certificates, passports (if you decide not to keep them with you), baptismal or similar documents; marriage and divorce papers; stock certificates and bonds; abstracts of title, deeds, and deeds of trust to real property; deeds to burial plots and burial instructions; military discharge papers; and wills, trust documents, and insurance policies; as well as written appraisals of your valuable personal belongings. Give copies of these documents and the duplicate key to your safe deposit box to your lawyer, a close relative, or trusted friend who lives in an area away from the fire danger. If you have appointed someone as your attorney-in-fact or agent by means of a power of attorney, a medical durable power of attorney, or a living will, that person should probably keep the original document in a safe place accessible to them. Ask your lawyer if you are not sure where to keep a particular document.

- **Keep receipts and record serial numbers.** If you can't prove exactly what you owned, when you bought it, and what it cost, you may not like what the insurance company offers as a "replacement." This information should also go in the safe-deposit box, with a copy in another safe place. Copies are fine for most insurance company purposes.

- **Take photos or videos of everything.** You will need proof that you actually owned something and be able to show its condition in order to make a claim for its loss or damage. Photos or videos of your property are a good way to do this. They should include complete views, both inside and out, of your home and other structures on your property, along with everything in them. Be sure to get close-ups of valuable and one-of-a-kind items, such as jewelry and antiques, as well as your cars and other motorized vehicles. Keep these photos and videos in your safe-deposit box. You may also want to keep a second set at a friend's house or at your place of business, if it is outside the fire-risk area. Your insurance company may, or may not, want copies.

- **Get appraisals of valuable items.** Insurance policies typically have limits on some insured items such as jewelry, antiques, silver, gold, collectibles, guns, electronics, and computers. If a fire destroys these items, you'll need to prove you owned them (see the preceding bullet), and you'll need to prove their value with an appraisal. If the value of any of these items or class of items exceeds the policy limits you have three choices: (1) keep them in a safe-deposit box (not always possible or practical); (2) take your chances that nothing will happen to them; or (3) buy appropriate "riders" (special insurance coverages) to protect them against loss or damage.

- **Make sure your insurance policies are adequate and paid on time.** If you don't have "full replacement coverage," you should. More on this in Chapter 5: Insurance.
- **Prepare "evacuation lists" and "evacuation boxes."** The details of what you should include and take with you are covered in several parts of this chapter below.

IF YOU ARE ON NOTICE OF A POTENTIAL EVACUATION ORDER

People must stay alert whenever wildfires are active in their region. Usually, you will have ample notice before you have to evacuate. If at all possible, fire authorities will try to give advance warning that an evacuation may be ordered and to "stay tuned" for further developments. Many residents of the wildland/urban interface spend much of each fire season prepared to evacuate on a few hours notice. This section contains useful checklists and steps to take if wildland fires are burning in your area or if a wildfire is actually approaching your property.

What Fire Authorities Want You To Do

Mark Your House Clearly

If you haven't already done so, put your house number in several places—on the house itself, at the entrance to your driveway, and on the road immediately in front of your house. Use fire-resistant materials to make these signs and secure them so they won't fall over or blow away. Spray paint on a large rock or a brick wall will also work if you have little time to prepare.

Remove Any Obstacles

Make your house as accessible as possible for firefighters. Even though you may leave in a rush, it is important that toys, bicycles, trailers, firewood, trash, and other impediments be moved out of the pathways to your house. This means they should not be in the driveway, around any entrances, near the water spigots, or anywhere in your defensible space.

Make Your House Easy to Find

If you are told to evacuate, leave the lights on to make your house easier for firefighters to spot in smoke and at night. This also shows that the power is working. Why is this important? If you are on a well, working lights are a pretty good indicator your well is working, and that means the firefighters can use your water spigots and hoses.

Escape and Evacuation Planning

The most important thing you can do is to make sure your family and guests know what to do in the event that a wildfire threatens your home. This means developing and practicing escape and evacuation plans.

First, everyone needs to know how to get out of the house if it catches fire, or if outside doors are blocked by fire or some other hazard. For example, climbing out a ground-floor window may be one means of escape. If your home is built on a slope or has more than one level, you should have rope ladders in, or easily accessible from, each upstairs room.

Second, everyone needs to know what to take with them and what to leave behind. If you have no warning and suddenly your life is in danger from a looming fire, *don't take anything* with you. Just get yourself and your loved ones out of harm's way. If you do have sufficient notice and time to prepare, there a several things you should be sure to take with you; these are explained in greater detail in the discussion of "Evacuation Lists" later in this chapter.

Third, everyone needs to know where to go once outside the house. If your house is the only thing burning, 50 to 100 feet may be the distance you have to go to reach safety. However, if a wildfire is threatening your neighborhood, 5 to 10 miles may be the distance you have to go to reach safety. Accordingly, your family evacuation plans should include at least two escape routes out of your neighborhood, along roads you are reasonably certain will be open in an emergency. Your family should also know where to gather initially just outside your house so that you can count noses. When everyone is accounted for, you can all get in your car and drive to safety together. You should also have a designated meeting spot at a safe location outside the fire danger area, in case you become separated or in case some family members are not at home when the evacuation takes place. Most importantly, know your escape routes and make sure they are open before proceeding.

Children and Other Family Members

Perhaps it goes without saying, but your children will want to know what is going on. They are likely to be frightened, so they need to know what plans you have made, where their things will be, and who might be taking care of them or picking them up if you can't. Make sure you go over basic safety, where to go if they are separated from you, whom to call if they can't reach you and have to evacuate from school or a friend's house, and how to "stop, drop, and roll" in case they have to flee through flames. Children, and your entire family, also need to practice using any emergency evacuation routes and safety equipment (such as ladders to get out of upstairs rooms).

Always know where your children and any disabled or elderly family members are.

Clothing

If you are under a potential evacuation notice, get in the habit of wearing or carrying protective clothing. You might have everyone keep a daypack with them containing these items. Protective clothing includes sturdy shoes or boots, non-flammable clothing (natural fibers, not synthetics), long pants and sleeves (even in summer), and carrying a handkerchief and gloves when you're outside. When you're inside, keep the stuff you're not wearing at the door in your daypack so you won't forget it.

Dogs, Cats, Horses, and Other Pets

It's usually not practical to take your pets with you every time you leave the house. You may be able to take your dogs with you on short outings and leave them in the car (assuming they have water and adequate ventilation)—they'll probably be perfectly happy. But cats and most other household animals aren't so easy to deal with. If you keep all your pets in a single room when you're away, you can show your neighbors which room they'll be in and which window or door to use (or smash) in order to get them out, if necessary. Otherwise, you'll need to make accommodations for your pets during the fire season outside the potential fire zone. Even if no fire is looming, make sure a neighbor knows when you'll be gone for a day or more and let them know about your stay-at-home pets.

If you own larger domestic animals—such as horses, especially, or even more exotic animals like pygmy goats, potbellied pigs, or llamas—you probably view them as pets or even as members of your family. You'll want to make arrangements to take them with you if you are forced to evacuate your home. That means having available a horse trailer or other hauling vehicle large enough to accommodate all of your large animals; stocked with feed, tack and accessories, grooming supplies, and medications; and ready to load up quickly when you get the order to evacuate. Keep in mind that hauling a trailer behind your car or pickup may make it more difficult to evacuate the area safely, so allow yourself plenty of time. And know where you're going. Have a pre-arranged boarding facility or stable set up, notified, and ready to accept your animals when you arrive. Your hotel probably won't let you keep them in your room.

Fire Information

Because your life and your property may depend on it, stay informed of the latest developments with regard to existing wildland fires in your region. Wildland fires can advance several miles in a single day and may change direction without warning. So it's a good idea to keep a radio or TV tuned to a local station that is giving fire and evacuation updates. Your local sheriff, police, or fire department can tell you the best channel or frequency for fire information.

WHAT TO TAKE WITH YOU

"When we smelled smoke Saturday afternoon, I called a neighbor who told us there was no need to worry, that the fire was 25 to 30 miles to the South. By mid-morning the next day the smell of smoke got stronger and soon the smoke was visible. The neighbor said there was still no danger so we drove to Bailey for lunch. From there we could see the smoke billowing thousands of feet in the sky—black and terrifying. We immediately drove back to the cabin and watched the smoke the rest of the afternoon. We don't have a television at the cabin and the radio didn't have any news, so we used the telephone to get updates. The smoke was unbelievable, angry, and wildly colorful. I gathered a few things and took them to our pump house, which is underground, but trying to decide or choose what to save was overwhelming."

— *Candace Boyle*

Evacuation Lists

You should prepare written lists of the things you'll want to have with you during an evacuation and things you need to do just before you leave. Make these lists well ahead of time so you won't forget any important items. Your lists should include: (1) an "Emergency Telephone List"; (2) a "Last-Minute Grab List"; (3) a "Pre-Packed Items List"; and (4) a "To Do Before Leaving List." These lists and what to put on them are explained in greater detail below and are reproduced at the back of the book for easy access and copying.[23]

---

[23] The American Red Cross also has a wealth of information posted on their "Family Disaster Planning" website at http://www.redcross.org/services/disaster/beprepared/familyplan.html.

Post several copies of these lists where everyone in your house can find them—perhaps on the refrigerator door and by your main exit doors. You might also want to do a "dry run" by packing all the items on your lists into "evacuation boxes" and then loading them into your car, trailer, or other vehicle to make sure everything fits. If what you want to take with you doesn't fit, you'll be forced to leave some things behind when it comes time to leave. So, you may want to consider renting a storage locker in a more urbanized area ahead of time and keeping large or bulky items you don't need on a daily basis there for the duration of the fire season. If you have heirlooms or antique furniture you want to protect, consider "lending" them to a family member or close friend who lives outside the potential fire area until fire season is over. Remember, you may not have much time to pack up and get out, so it pays to move your larger valuables out of harm's way ahead of time.

The "Emergency Telephone List"

Keep a list of emergency telephone numbers in a conspicuous place in your home and in all of your vehicles. It is even a good idea to make additional copies of the list so that family members and visitors will each have one. Most cell phones can be programmed to store dozens, if not hundreds, of names and numbers, so take advantage of that feature if you have it. Your list might include the following numbers:

- 911, and instructions on when it is appropriate to use it.
- All responding fire departments in your area (there may be more than one).
- Local law enforcement and emergency medical dispatchers.
- The state patrol number for information on road conditions and possible closures.
- Your children's schools and daycare facilities.
- People who are authorized to pick up your children if you are unable to.
- Relatives, close friends, and others outside the potential fire zone who should be notified of your whereabouts.
- Your neighbors.
- Your business office or employer.
- Your physicians, pharmacies, veterinarians, and medical facilities.
- Boarding facilities for horses and other large domestic animals.

You may already have most of these numbers in your address book, day planner, cell phone, or personal digital assistant (PDA), which you'll also want to have with you during an evacuation. But everyone in your family should have a copy of your Emergency Telephone List.

The "Last-Minute Grab List"

These items are things you use or need to have available on a day-to-day basis, so it doesn't make sense to pack them up weeks in advance of a possible evacuation. However, you should consider keeping this list and a large plastic storage container (evacuation box) by your door or perhaps in your front hall closet. The list will remind you of everything you need to grab; the box provides a convenient receptacle into which you can toss everything quickly and carry it out with you as you leave. Your Last-Minute Grab List might include the following items:

- All *medications* and medical supplies for everyone in your family and your pets.
- Your driver's license and passport.
- Any cash, checkbooks, credit and gift cards, calling cards, and similar items.
- Your address book, day planner, or PDA, and a copy of your Emergency Telephone List.
- Portable electronic devices, such as your cell phone *and charger,* and your laptop computer and peripherals. If you have backed up your home computer files and operating system, take those disks, too.
- Glasses and contact lenses, with cleaning and storing supplies.
- A special blanket or stuffed animal that your child will want at bed-time.
- The children's school textbooks and notebooks.
- Jewelry that can't be replaced (take only the "real" stuff).
- Cameras, exposed but undeveloped film, and video equipment.
- Pet stuff, such as leashes, medicine, and food.

The "Pre-Packed Items List"

Your actual choices will vary according to where you'll stay or how long you'll be gone when you evacuate, but the following Pre-Packed Items List is a good start on planning for the things you'll need. Give some thought to what additional items should be on your list, according to your particular needs. It's a good idea to have all of your pre-packed items boxed up and stacked in a closet by the door or next to your car, preferably in easy-to-carry plastic storage containers (evacuation boxes) with lids. Some people even spend the whole fire season with all these items in their cars.

- This book! (Chapter 4: Recovery contains useful information on what to do *after* the fire.)
- Copies of all prescriptions and medical records for you family, including a list of all prescription numbers, where they are on file, and the pharmacy phone numbers, especially those obtained by mail-order.
- Additional copies of your Emergency Telephone List.
- Insurance information, including your policy numbers, agents' names and phone numbers, and insurance cards.
- Photocopies of all the documents in your safe-deposit box.
- Books and toys for the children.
- Several days' worth of clothing and a jacket or sweater for each person.
- Toiletries and other personal care items.
- Non-perishable snacks and bottled water.
- Pet carriers, crates, toys, extra food, and feeding dishes, as well as feed, tack and accessories, grooming supplies, and medications for horses and other large domestic animals.
- Copies of your children's school records, scrapbooks, and any collections of school projects, awards, and achievements you've been saving for them.
- Other items you can easily carry and would hate to lose, such as family photos, mementos, and heirlooms, as well as home movies or video tapes of children's birthdays, graduations, weddings, and other family events.

Where you're likely to stay during an evacuation may help to determine the items you will need and want to take with you. Here is a list of some things to consider for your *"Pre-Packed List"* according to where you stay:

- If you are going to stay with a close friend or family member, you will want all of your evacuation list items, but you may not need to bring food unless you have special dietary requirements.
- If you will be staying in a community shelter, this option requires some additional planning on your part. For example, you'll want to bring lots of toys for the children, things to entertain the adults, and your own linens, pillows, and towels, as well as shower shoes to wear in community showers. Extra linens are useful for creating privacy screens. You may also want to bring your own sleeping bags and mats

since the cots may not be very comfortable. Whatever snacks and special foods you can bring with you will help add some variety to the cafeteria-style food.

- If you go to an extended stay hotel or similar apartment-like place, you may also want to add whatever food you can reasonably manage to bring from your pantry, since you will have cooking facilities. Books, games, videos and other forms of entertainment will help you pass the time. Since you'll be in your "own place," you could bring more clothing if time permits.

The "To Do Before Leaving List"

The fourth evacuation list mentioned above is the "To Do Before Leaving List." With several hours' notice, you may have time to accomplish a number of additional tasks that will help to protect your home and make the firefighters' jobs easier. You may want to copy this list (reproduced in the back of the book) or make your own list and add the things that are specific to your situation. Keep it with your other lists as a reminder.

- Shut off the gas at the meter or propane tank.
- Leave your outside (and inside) lights on to help firefighters find your house through smoke and darkness. This also lets firefighters know there is power to the house; this means your well or water system should still be working.
- Close windows, shutters, vents, fire resistant or heavy drapes, blinds, and doors to help block radiant heat. Locking the doors and windows is up to you. Leaving them unlocked gives firefighters easy access without having to break in, but obviously it also leaves your house unsecured.
- Take down sheer curtains; they are very flammable.
- Open the fireplace damper to allow hot air to vent out of the house. Your chimney should already have a spark arrestor installed, but as an added precaution, position the fireplace screen over the hearth to prevent sparks and embers from being blown in.
- Move furniture to the center of each room. Try to get flammable items away from doors and windows, especially large glass doors and windows.
- Wet down any plants or trees within 15-20 feet of your house and other buildings.

- Put a sprinkler on your roof (unless it's composed of non-combustible materials, such as tile or metal). Wet the roof down before you leave.
- Put garden hoses and buckets filled with water around the house. Firefighters can use these to put out spot fires and flare-ups, if necessary.
- Position a metal ladder next to your house for firefighters to use to access your roof.
- Seal outside vents with heavy plywood or non-flammable barriers, such as rocks or gravel.

Author's Note:

My friend Barbara has witnessed three fires. Her suggestions are included in the four evacuation lists described above. In addition, here are some of the things she and I have learned from our various experiences and plans we have with our neighbors to coordinate future evacuations.

1. Get to know your neighbors! The residents on Barbara's cul-de-sac had a meeting to set up a telephone tree for evacuation notification. That's when they found out that one woman didn't drive, even though she and her husband owned two cars. Without this meeting, no one would have known. Now they have an evacuation plan for their road that includes making sure this woman is not left behind if her husband is away. They also have plans for rescuing pets and pre-packed belongings if someone isn't home when the evacuation order comes. We did the same thing on my road since many of us work at home, but others don't, and we have pets, disabled family members, and elderly parents to consider.

2. Call your local fire department to ask about how to build a notification tree and for any recommendations about evacuation and what to take.

3. If fire authorities have instructed you to prepare for a possible evacuation, make sure several neighbors know whenever you'll be away from your home for any length of time. They can then take care of whatever they have time for at your house, but only if they know you're not there. Consider giving them keys so they don't have to break in.

4. Make sure a family member outside the area, one or two trusted neighbors, and at least one friend *who lives outside your neighborhood* have permission on file at your children's schools and daycare

facilities so they can pick up your kids during an evacuation if you're not around or can't get there. At least they'll be with someone you (and they) know.

5. Keep an overnight bag packed, with two changes of clothes for each person, basic personal care items, all prescriptions and medications, and other immediate, hard-to-replace items. You may want to keep this bag in your car. Don't forget to pack one for the pets.

6. Consider having the children take one overnight bag to school and pack a second one to keep in the family car. Children in many neighborhoods go to schools that are on the other side of potential road closures from their homes, meaning they could get separated from their parents. So it's nice for them to have some of their own things when they get to a shelter or friend's house.

7. Pets feel threatened by unknown situations and will sense your concern and fears about approaching wildfires. Do your best to calm them. The last thing you want is for your pet to run off during this time. So be sure never to take a dog outside unless it's on a leash. Keep cats and other pets inside until you are ready to evacuate, and then transport them in pet carriers.

8. Store any valuables you don't need day-to-day in a safe place—a safe-deposit box in the closest big town or city, preferably—or with a trusted friend, or at your office (if it's not in the fire risk area).

9. If you travel a lot on business, it's a good idea to keep your passport on your person throughout the fire season.

10. Get into the habit of keeping interior doors closed so you won't forget to close them if you have to flee. Closed doors help to reduce airflow and drafts that spread a fire.

11. Always keep your escape vehicles backed into your garage or driveway, pointed the way you'll need to go when you evacuate. If an evacuation is imminent, leave the keys in the ignition, roll up the windows, and leave the doors unlocked. Keep the gas tanks at least half full throughout the fire season. Keep a first aid kit, warning triangle, and fire extinguisher in each car.

12. If you are leaving under an evacuation order, disconnect your automatic garage door opener by pulling the emergency release handle; the power may go out in your area, rendering the door nearly impossible to open from the outside if you don't. Leave all doors and windows in your house closed, but preferably unlocked for firefighters.

13.   Connect garden hoses to outside spigots. If you have more than one spigot, you should have a hose for each one. A brightly colored hose is easier for firefighters to spot than a green or brown one.

14.   If you have a hot tub, fish pond, or pool, keep it full. Firefighters may be able to use the water.

15.   Check to make sure that your homeowner's or renter's insurance covers additional living expenses in an extended-stay hotel in case fire authorities order you to evacuate your primary residence. Prearrange your shelter so you know you are going to a place that accommodates your particular needs, including pets, if necessary.

16.   Make sure you have your cell phone with you at all times. Keep the phone charged. A car charger will come in very handy. Make sure several neighbors, your children's school or daycare center, and your children all have your cell number.

CONCLUSION

Hopefully, the evacuation lists and the preparation tips discussed in this chapter have been useful and informative. Their purpose is to help you and your family *live with wildfires.* But they won't help you unless you actually take the time to think through your particular situation, create your own evacuation lists, and take the recommended steps to protect your family and your property. Being prepared in the advance will also lessen your stress level during an evacuation. Now is the time to start, before wildfires arise and threaten your neighborhood.

Chapter 3: Evacuation

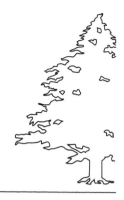

Lucky to be Alive

A Personal Account of the Hayman Fire by Kurt Hanes[24]

"I sat on the front porch of our historic 57-year-old cabin tucked away on the Tarryall River during a still sunny summer Saturday evening. My eleven-year-old son was down fishing the river with the small crew of friends he had joining us for the weekend. Their fathers had come along as well, so we were wrapping up the first full day of a true "Boys Weekend" at the cabin. Another father and I had just settled into our patio chairs to absorb a panoramic view from the Tarryall Mountains southward towards Pilot Peak when I saw the first billow of dark black smoke over the ridge. Amazed at the short proximity we jumped in his Suburban and headed toward the ever-growing black plume.

"We arrived at the approaching flames and towering lively smoke by means of a short Jeep road and quickly realized it was already out of control with less than 10 minutes of burn time. We turned around and passed a pickup truck with the first four fire fighters arriving on scene. They rejected our offer to help and suggested leaving the area as the most appropriate action. Headed back to the cabin we noted a larger fire truck filling its hold from the Platte River....

"We were back at the cabin within 15 minutes of our first sighting. The boys and other fathers had abandoned their fishing in lieu of the nearby spectacle. Slurry bombers and lead planes appeared out of nowhere as if they were laying in wait. ...Certainly with fire crews and trucks on scene, natural barriers such as roads and rivers and three slurry bombers dropping their loads of red retardant they had caught the Hayman Fire in time.

"The black smoke turned to gray exhibiting an intense internal red glow. We watched for the smoke plume to dissipate and the flurry of

---

[24] Reprinted by permission. For pictures and additional information, visit Kirk Hanes's website at http://www.kirkhanes.com/.

bombers to victoriously fly away as a lone county sheriff drove up our dirt road.

"Within an hour we had packed not only our weekend gear but most of the keepsake memorabilia that had gathered over 57 years. Stuffed trophy fish, black and white photos in real silver picture frames and LP records of barbershop quartets were leaving the cabin for the first and perhaps last time. Boys Weekend had indeed turned into the Hayman Fire, although we didn't know it at the time.

"Weeks would pass with no end to the destructive rage in sight. Thousands would follow our evacuation lead with varying outcomes. The fire and accompanying smoke was now approaching my home thirty miles from the ignition point.... [and I had to] re-gather the keepsake memorabilia of the cabin combined with household goods and move them once again. My family joked the fire didn't want the cabin, just the stuffed trophy fish."

–Kirk Hanes, Hayman Fire (2002)

The experience of Kirk Hanes and his friends is only one of thousands where quick response by law enforcement and common sense and planning has saved lives when confronted with wildfires. This chapter will help you plan what to do if you are ordered to evacuate based on how much time you might have to leave, what to do before you leave, how you can help firefighters protect your house and themselves when they arrive, places you might stay and how to monitor the fire. The important thing to remember is that regardless of how much time you think you might have, and how much you would like to take or do, your safety and peace of mind are the most important things—and you should leave as quickly as possible.

WHEN THE EVACUATION IS ORDERED

You may have minutes to escape or days to prepare to leave. Below are suggestions about how to use the lists and preparations you have made to evacuate safely and to be more comfortable while you wait to return.

Your Escape Route

If fire authorities tell you to evacuate immediately and to use an escape route that is different from the one you planned, or to go in a direction you hadn't planned, it's a good idea to follow their instructions. This may separate you from your family and friends if they are in a different area at the time, but that can't be helped. If you try to go to a place or in a direction contrary to that specified by the authorities, you may put your-self at risk and may impede or prevent the escape, rescue, or firefighting efforts of others. Besides, unknown to you, your planned escape route might be blocked by emergency vehicles and personnel, or by the fire itself. Drive cautiously and keep an eye out for firefighters and fire trucks that might be approaching from the opposite direction and possibly obscured by smoke and terrain.

No Warning

If the fire is imminent and you get no warning, take your family and your household pets, and *get out.* Your house and your personal belongings are not worth the risk of death or serious injury to you or your family.

30 Minutes' Notice to Leave Your Home

If you are ordered to evacuate in 30 minutes or less, try to gather and take with you as many of the items on your Last-Minute Grab List as pos-sible. To the extent it is still safe to do so, you can load up your car with additional evacuation boxes you already have prepared from your Pre-Packed Items List. Don't try to take anything else; insurance will replace most things, and 30 minutes isn't very long. If there is only one road in and out of your property or subdivision, you and your family *really* need to get out as fast as possible, or you may not get out at all.

Leave your house lights on, unless fire authorities have advised you to do otherwise. Turn off the propane or natural gas. If you have a well, leave the electricity to the pump on; firefighters may need to use the water. Then drive quickly, but cautiously, out of the fire danger area.

Several Hours' Notice to Leave Your Home

With this much notice before you have to evacuate, you should have time to collect everything on your Last-Minute Grab List and to load all of the evacuation boxes from your Pre-Packed Items List into your car. To the extent it is safe to do so in whatever time you have remaining, you can use more storage containers and even suitcases to pack up additional items, such as:

- Several more changes of clothing.
- Food to supplement what you'll get at the shelters and to meet special dietary needs.
- Additional photos, artwork, and collectibles.

- Small antiques and other irreplaceable items, such as family china and silver.
- Extra children's books and toys.

Of course, these additional items still have to fit into your car or trailer with your family and the necessary items you've already loaded into it. So, try not to get carried away!

With several hours' notice, you may have time to accomplish a number of additional tasks on the "To Do Before Leaving List" you created after reading the last chapter. These are the things that will help firefighters find and protect your home, make the firefighters' jobs easier, and possibly minimize the damage by the fire. There may be other things you have added to your list that you have time to do that are specific to your situation, like filling horse troughs, opening automatic gates, or removing fencing that might be confining to animals. So, after your car is packed, your family and pets are all accounted for, and you're ready to go, you may wish to do as many of those items as can safely be accomplished in the time you have remaining. (But be sure to leave yourself a cushion: Don't wait until the last minute to evacuate.)

IF YOU ARE SURROUNDED OR TRAPPED BY A FIRE[25]

If fire is closing in around you but you can still leave your house safely, or you are ordered to evacuate the area immediately, turn the outside lights on, and then *get out*. Leave the door you use to exit unlocked for firefighters. If the worst happens and you are unable to evacuate the fire area, the following safety tips may help save your life:

If You Are Trapped in Your Home

Call 911, give your address, and say you are trapped in your home surrounded by a wildfire. To the extent it is safe to do so, follow the steps outlined above in the "To Do Before Leaving List" section. Then take cover in the inner-most part of your house on the ground floor. If you have a cell phone, *keep it with you* (don't forget the charger). Stuff wet towels along door cracks to keep out the smoke and fumes. Avoid putting a wet cloth over your face; the heat of the fire can superheat the water and sear

---

[25] Adapted from "Safety Tips During a Wildfire" (August 25, 1999) by the California Department of Forestry and Fire Protection.

your nasal passages, throat, and lungs. A dry cloth is safer. Once you have reached the safest interior location and closed the doors, leave them closed and wait for the fire to pass or for emergency personnel to arrive.

If You Are Trapped in Your Car

If you can see through the flames and smoke to an escape route, drive through them as quickly and safely as possible. Turn your headlights on so others can see your car approaching. If it is too smoky to see, pull off the road into the biggest clearing you can find. Roll up the windows. Leave the engine running with sufficient RPMs to keep it from stalling. You may keep the air conditioner on, but be sure to set the AC on "Recirculate" to close the outside vents and keep out smoke and fumes. Get down on the floor of the vehicle, on the side of the car that is farthest away from the approaching fire. Cover yourself with anything you have in the car—blankets, coats, extra clothes. Place a dry cloth over your mouth and nose and breathe as shallowly as possible. Automobile gas tanks rarely explode, but cars do (and probably will, in this case) catch fire. Remember, no matter how bad it gets inside the car, it is probably worse outside and the car will provide some protection. So stay put until you are sure the fire has passed. Then quickly and carefully get out of the car; it may still be burning. Cover your face and hair, if possible, and again, take only shallow breaths. Even if you think you have no flames on your body and clothes, once you are clear of the flames, do the "stop, drop, and roll."

If You Are Trapped on Foot

Try to get downhill from the fire since convection (rising hot air) causes fire to burn faster uphill. However, keep an eye out for rocks and burning debris rolling down the hill. Get as low as possible—in a dry creek bed or depression, for example. If you get low enough, a crown fire may simply pass over you. Of course, if you can make it to a stream, lake, or river, submerge yourself in the water. Another option is to find a wide open area with little vegetation and rocky soil. You may also find some protection behind or under a rocky outcropping. But remember, the flames may heat the rock to extreme temperatures and you can be burned by radiant heat from the fire as well, so a shallow outcropping may not provide sufficient protection. If the approaching wall of flames is thin enough that you can see to the other side, you may—*as a last resort*—be able to run through the fire to an area of relative safety. *Don't inhale!* Use a dry cloth to cover your face and hair, if possible. Again, once you are clear of the flames, do the "stop, drop, and roll."

If you can't get to water, another relatively safe spot to go is an area that has already burned *completely,* since a *reburn* of the area is unlikely. Be cautious, however: The ground will be hot, and there may be sinkholes over burning tree roots. Don't touch any tree stumps since they may still be burning inside or below ground level. In fact, stay clear of all standing trees in a burned area, whether they are severely burned or are still green and appear not to be burned at all. Firefighters call these trees "snags." The fire and the microclimate it created may have weakened the roots, branches, and support structure of these trees. Trees in this condition have been known to topple over, slough off large pieces of bark, or drop weakened branches, hitting people and causing severe injuries and even death.

WHERE YOU WILL GO IF YOU ARE ORDERED TO EVACUATE

Where you're likely to stay during an evacuation may help to determine the items you will need and want to take with you. If fire authorities order an evacuation, you will likely end up in one of three places:

- **The home of a close friend or relative.** For many people, especially those without insurance and those unfamiliar with what their insurance policy covers, this is the most frequently made choice. Again, you will want all of your evacuation list items, but you may not need to bring food unless you have special dietary requirements.

- **A community shelter.** This is probably where you will stay if you feel the need to be close to your neighborhood (to monitor the fire situation and your property), or your children need to be close to their school, or you have nowhere else to go. Emergency aid organizations and community volunteers try to make the shelters as comfortable and homey as possible. But with hundreds of people living together under one roof, perhaps in a gymnasium or church social hall, there is little privacy and the food is typically cafeteria style. So this option requires some additional planning on your part. For example, you'll want to bring lots of toys for the children, stuff to entertain the adults, and your own linens, pillows, and towels, as well as shower shoes to wear in the community showers. You may also want to bring your own sleeping bags and mats since the cots may not be very comfortable. Whatever snacks and special foods you can bring with you will help add some variety to the cafeteria-style food. And extra linens are useful for creating privacy screens.

- **An extended stay hotel** or similar apartment-like place. If you have homeowner's, renter's, or condominium insurance, your policy should cover additional living expenses when you are ordered out of your property due to an approaching fire. You will want to bring along as many of the items from your evacuation lists as possible. You may also want to add whatever food you can reasonably manage to bring from your pantry, since you will have cooking facilities. Books, games, videos and other forms of entertainment will help you pass the time. Since you'll be in your "own place," you could bring more clothing if time permits. On each trip back to your home (assuming you are allowed to return briefly during the evacuation period), you can retrieve more of your belongings. Make a list before each trip so you can pull out the additional items as quickly as possible.

KEEPING INFORMED

"The Hayman, Buffalo Creek, and High Meadow fires all came within two to three miles of our properties. During the Hayman fire, we got a reverse 911 call just after we had gone to bed at about 9:30. Family and neighbors said the danger was still far away but we decided to play it safe and head for Denver. As we drove down our road we could see the red glow of the fire in the distance. The most difficult part of any of the fires that have surrounded us has been not being there, the uncertainty, helplessness, and lack of accurate information about the exact movement of the fire."

—Candace Boyle

Dealing with the uncertainties of what is happening to your property, and what it will cost if you have substantial losses are probably some of the most difficult things to face. But, there is no need for you to face these things alone. Here are some ideas to help you stay in touch with what is happening and where to get help right away.

1. Immediately after you have been forced from your property because of the fire and have reached an evacuation center or other safe location—even before you know whether your property has been damaged or destroyed—you should notify your insurance company of your situation and potential loss. It is best to notify *all* of your insurance carriers (e.g., property, auto, liability, business) since any damages you incur may be covered under more than one policy.

Since your policy probably covers your expenses while you are out of your home, verifying this with your agent will immediately help dissipate your fears about the cost keeping a roof over your head. If you had to leave a car in the garage, your agent can remind you that fire damage is covered. Your agent may also be able to suggest places to stay, and have suggestions for coping with the time you are out of your home. Remember, they are experts in these types of situations.

2. Contact the law enforcement, fire authority, or forest service agency that ordered the evacuation to find out how and where you can get current information on where the fire is going and when it might be safe to return to your property. Be aware that these organizations may have a limited amount of time to talk to you, due to the high volume of calls and the need to deal with the fire itself.

3. Listen to local television and radio news reports.

4. News stations and the forest service usually have web sites with lots of links that cover wildfires in more detail than is possible to show during regular news hours. These sites have more current information and detailed maps that show the fire's location and direction. Some even have chat rooms so that you can "talk" to those who may be in visual range of your area.

5. Community shelters where those who have been evacuated are staying are good sources of information. The fire organizations hold regular updates and you may get the opportunity to ask about your specific property.

CONCLUSION

Leaving a home or cabin that is dear to your heart may be one of the hardest things you have to do. The steps mentioned in this chapter can help you improve your chances to get out safely, assist firefighters in saving your property, and make your stay outside your home more comfortable. But, the one thing to keep in mind is that the safety of your person, your family, and your pets should always come first. Use good common sense to evaluate your time and evacuate as soon as possible.

Chapter 4:
Recovery

Recovery is perhaps the most overlooked aspect of living with wild-fires, from both a practical and an emotional standpoint. But if a wildfire has destroyed or seriously damaged your home, knowing what to do and organizing your own recovery plan can help you deal with the heartache—and save you some headaches as well. This chapter contains useful information about what to do first, how to return safely to your property, what your responsibilities are after you've suffered a loss, and how best to deal with repairs, rebuilding, and the emotional trauma you're likely to experience, as well as government agencies and relief organizations that can provide assistance.

Once the fire danger has passed, you will be faced with the emotional experience of returning to your home or business. You may not know if anything is damaged or your home is still standing. That first glimpse will be one of the most telling moments of your life. You may have seen news reports of people returning to their homes and finding nothing but a mound of burned rubble, a child's bed frame, a shattered chimney, or a twisted scrap of metal that used to be a stove. Fortunately, there are many more happy stories about people returning to find their homes intact, even in cases where the surrounding trees have burned to the ground. In either situation, this chapter will help you deal with the additional threat of damage to your property from the floods, landslides, and mudflows that often follow in the aftermath of a fire. Whatever your personal experience is, you'll want to deal with it head on.

> "The Hayman fire was the largest recorded fire in Colorado history. It burned over 137,000 acres consuming 600 structures of which 133 were homes or cabins. *Mine was not one of them.* I have since driven and hiked through the burn area as well as flown over it in a private plane. As a Colorado native and avid outdoor enthusiast, I personally contemplate the outcome and my position on the event.

The so-called destruction of Mother Nature is overwhelming. The view from the air places keen measurement on the magnitude of the event. The most amazing discovery is the randomness of what burns and what stays. The number of blackened tree trunks is countless as ash and soot clog drainage pathways. The ground is black as well and void of most features. We hear of the potential for mudslides, the danger of falling trees and an enormous effort to rehabilitate an area destroyed by one of our own, mankind.

As one of mankind I have always been amazed at our need to get involved with everything and fix it if broken just as we are doing with the Hayman Fire. Clearly the loss of homes is devastating to those involved. However, I came to the realization that if the cabin did burn I could easily accept that reality.

Who is to know the fire would not have started by lightening or another force of nature the following day, week or month. Mother Nature has been destroying herself and rebuilding herself for millions of years. I argue she knows what she is doing as evidenced by the beauty she surrounds us with.

So the forest is black and we feel the need to intervene and fix it up just as quickly as possible. ... Why is non-involvement not a solution? The Hayman Fire is just another small point in the time continuum just like Boys Weekend and the natural rehabilitation of the burn area. Perhaps we won't be alive to see how much more fully the forest will recover given the appropriate amount of time, but someone will.

... So I now sit on the porch of the cabin in the same patio chair. The barbershop quartet can be heard playing on the record player inside and the stuffed trophy fish hangs yet again above the mantle. The panoramic view contains entire hillsides of blackened lifeless toothpicks spotted with areas of green healthy forest. How lucky I am to be alive and experience one of Mother Nature's resurrections."

— *Kirk Hanes, Hayman Fire (2002)*[26]

---

[26] Reprinted by permission. For pictures and additional information, visit Kirk Hanes's website at http://www.kirkhanes.com/.

THE FIRST STEPS TO RECOVERY

Once you are told you can safely return to your property, there are several things you need to take care of in fairly short order, and lots of precautions you should take. Here is a list of the things you should *do first* to get started on the road to recovery. The necessity of each step and the sequence of steps may vary somewhat depending on your personal situation.

1. Listen to news reports or contact the law enforcement or fire authority that ordered the evacuation to find out when it is safe to return to your property. The same authority can tell you when it is safe to begin the clean-up process.

2. Be sure to review the "Safely Returning to Your Property" section below before you venture into the fire area and onto your property.

3. **Let your insurance company know you have been given permission to return to your property and ask to have their adjuster meet you there.** If the insurance company is going to pay for it, you'll need their approval to arrange for clean-up, demolition, repair, and rebuilding; so it's best not to begin these activities without their involvement. (**Note:** Be sure to keep receipts for any emergency repairs you have to perform before you are able to meet with the adjuster.)

4. You should already have "Before" photos or videos of your home and property. Now is the time to pull out that camera you saved from the fire and take photos or videos of everything again, damaged or not. Your "After" photos are very important. The insurance adjuster will also take photos, but no one knows your property as well as you do. Comparing your Before and After photos can help in resolving potential disagreements over the existence, prior condition, or subsequent damage to your property.

5. Once you have determined what work will be necessary to restore your property, you can make the arrangements for clean-up, demolition, repair, and rebuilding through your insurance company or through a company-approved vendor. It is probably a good idea to get the insurance company to issue its approval of the work and the vendors in writing.

6. If you do not have insurance, you will need to arrange and pay for the clean up, demolition, repair and rebuilding yourself. Finding an established company that specializes in the work you need done can be difficult. Reading "Choosing the Right Contractor" later in this chapter will help, and warns you about fly-by-night companies to avoid.

Later in this chapter there is information about the Federal Emergency Management Agency (FEMA), the American Red Cross, the Natural Resources Conservation Service, the Small Business Administration, and other agencies that may be able to offer financial aid and practical assistance. It is definitely to your advantage to contact as many of these agencies as possible. The worst they can do is tell you "no."

SAFELY RETURNING TO YOUR PROPERTY

"This had been a family's life only a couple days ago. All that was left were the hard parts: a sink, copper plumbing, steel food cans boiled to bursting. The windshield of a truck out front had melted and poured into the cab, dripping like candle wax."

—*"The Anatomy of Fire" by Craig Childs, High Country News, Vol. 34, No. 13 (July 8, 2002)*

Before you actually set foot on your property, there are a number of precautions you should take. Danger may still be present, even though the fire has been put out. If possible, you should take someone with you to survey the damage—for safety and moral support. Be careful where you step. Trees may be ready to topple and the ground may be soggy or unstable. Fire may linger in the roots of burned trees, and jarring the stumps could bring the fire to the surface. There may be *ash pits*—holes in the ground filled with hot ash where trees once stood. Ash pits are often hard to spot and are very dangerous. Even firefighters have been severely burned by accidentally stepping into them. That's one reason it's so important to wear sturdy boots and leather gloves.

Another thing to beware of is flare-ups. Piles of ash might be covering hot spots that could suddenly burst into flames without warning. This happened during the 1991 Oakland Hills fires. Even though all visible flames from a much smaller initial fire had been completely extinguished, the "Diablo" winds blew in and quickly fanned the few remaining hot spots into a conflagration that destroyed hundreds of homes and killed 25 people. The property damage was well over $1 billion.

There may be any number of dangers to watch out for on your property. The stability of the remnants of your house or out buildings may be difficult to detect. Walls may be ready to topple, or ceilings and the roof ready to collapse, so you will need to be very careful if you enter any structures. It is possible there will be broken pipes that may be leaking natural or propane gas, and just one spark could set them ablaze. Inside electrical lines may be exposed or broken but hidden in the walls and ceilings. Every year, dozens of people are seriously injured or killed by inadvertently touching downed power lines. So, if you see a downed line, you should steer clear of the area and report it to the power company immediately. Even firefighters prefer to let the electrical experts fix these problems.

Finally, be aware that dangers you would never have imagined may also be present. Wildlife—even bears, wolves, and other large animals—may be lurking on the premises, frightened and looking for food. Looters are another unfortunate, but possible danger. It's best to assume anyone trespassing on your property is armed and does not have your best interests at heart. Therefore, it's wise never to confront a trespasser; leave immediately and call law enforcement for help.

What follows is a compilation of suggestions, hints, and safety tips from many sources. But no matter how careful you are, unexpected things can still happen and following the advice below in no way guarantees that you will be completely safe when entering a burned area or structure. If you are unsure about the safety of a particular location—*stay out!* Your local fire, sheriff, and police departments, state emergency management office, and regional FEMA office can also be valuable sources of information and safety tips.

Before You Return to Your Property

After the evacuation order has been lifted you should confirm with law enforcement or fire authorities that it is safe and permissible to enter the area where your property is located. Once you have the green light, proceed with caution. Take it slow and easy, remembering that you are probably experiencing a great deal of emotional stress. You can use the following checklist as a reminder of safety precautions:

• Always let someone know where you are going and when you plan to return. Ideally, you should bring someone along with you and let a third party know of your whereabouts. The safest thing to do is to have the person who comes with you stay in the car with a cell phone, and set a time for your return. If you are more than a few minutes late, the person in the car can assume you have run into trouble and call for help.

• Bring a first aid kit and plenty of drinking water with you. Take frequent breaks and keep yourself hydrated.

- It is a good idea to check with your local utility service providers before using or turning on any utilities such as gas, electric, or water. Damaged utility lines or connections can be dangerous or even re-ignite a fire. So, you really shouldn't flip a light switch, or reset a breaker, or even use the telephone until you've confirmed that your utilities are in good shape.

- Avoid driving too close to fire-damaged areas because the ground may be unstable and still hiding hot spots. Stay on established roads and park either on the road or on another hard, preferably gravel-covered surface, and turn off the engine. Your engine generates heat and could potentially ignite a fire, especially if you were to park on an area covered with dry grass or leaves. If the place you've chosen to park is soft and muddy, your car might get stuck and, if the fire flares up again, you could be trapped.

- You should wear long pants and a heavy long-sleeved shirt—both, preferably, made from 100% cotton, since synthetic materials are more flammable—sturdy boots, and leather work gloves. You should also bring along rubber gloves to protect your skin if you need to handle toxic substances or chemical irritants, as well as a painter's or surgical mask so you won't have to breathe ash and smoke. A hard hat and eye protection are also good safety measures. A burned area and damaged structures are like a construction zone, so it makes sense to wear the same kind of protective gear.

- You should bring a flashlight and extra batteries, since the power may not be restored yet, as well as a first aid kit to patch up minor injuries. You may also want to bring a hose with a sprayer attachment for both clean up and fire protection, assuming the water is on and safe to use. Alternatively, an A/B/C-labeled fire extinguisher would be a good thing to have if you need to extinguish anything that might still be smoldering.

- Make sure you *have* a cell phone with you, but keep it in the car or at least well away from any place that might potentially contain leaking natural gas or propane fumes. Cell phones can cause a spark or static discharge and ignite the fumes. Other electronic equipment, small appliances, and power tools have the same potential to ignite gas fumes, so plan to use them outside and well away from any other flammable materials.

- Perhaps it goes without saying that it's not a good idea to light any fires—including lanterns, camp stoves, matches, lighters, and cigarettes or other tobacco products anywhere near the property.

Approaching Structures

When you arrive at your property, look around the house and other structures for burning sparks or smoldering areas—*do not enter a structure if you see any evidence of potential or actual fire.* Go back to your car and notify fire authorities immediately—call 911. Then stay in your car, well away from the fire. Let the 911 operator know where you are and where the fire is; you may have to guide the firefighters to the flare-up or hot spot. Here are some additional safety tips to follow when approaching possibly damaged structures:

- If you don't see any sparks or hot spots in the area, inspect the outside of all structures for obvious damage and hazards such as leaning or sagging walls, falling materials, broken porches and decks, and out-of-kilter windows. If you're not sure whether something might be hazardous or that the building is safe to enter, don't. Is anything inside worth risking your life?

- Inspect the foundation—look for cracks, buckling, undermining (erosion), or other damage. Make sure roofs, decks, patios, and other overhangs are supported and not drooping or sagging. If you see any of these problems, *do not enter the building or go on the roof.*

- Once you're sure it is safe to go on the roof, you should inspect it to make sure there are no sparks or embers that could re-ignite. If you need to lean a ladder against the house, use caution since the house itself may be unstable. Have your companion hold the ladder as you climb up, since the ground around the house may also be unstable. If the ladder is metal (the best choice), make sure you are wearing thick gloves since the ladder could be warm.

- Try to open all of the external doors before you actually enter the building, not only to test the door themselves but also to ventilate the interior of lingering smoke or gas fumes. If any door sticks at the top, the building may be ready to cave in. Each time you open a door, you should stand clear of the door frame in case of falling debris.

- If it is safe to turn the water on and you can still use the outside spigots, consider watering down the area around your house—assuming there are no hanging wires or power lines that the water stream can reach.

- Anytime it appears unsafe to enter a structure, or you are not sure, you should stay clear of it and arrange for a building inspector to look at it. If you need a referral for an inspector, a Realtor® is a good source.

You may also want to check out the inspector with your local Better Business Bureau.

- If you left vehicles or other motorized equipment in the area, check for fire damage to them. Pay particular attention to the tires since the radiant heat from a fire will cause them to melt, or even explode, before the rest of the car shows any sign of damage. If you suspect your vehicles have been damaged and you carry comprehensive (or "other than collision") coverage, contact your auto insurance company immediately.

Entering Structures

As you prepare to enter a building, it's a good idea to take stock of your equipment and what you're wearing: hard hat, safety goggles, gloves, painter's mask, sleeves rolled down. You never know when something will drop off the ceiling or fall over and hit you, so it's better to be ready before you go in. You should also turn the gas off at the outside valve if you smell fumes or are at all concerned that gas may be leaking inside. If you're using a flashlight, it's best to turn it on before you enter the building. There's a slim chance the flashlight could generate a spark and ignite leaking gas or propane.

Once you enter the structure:

- Watch where you step and move slowly. You may encounter debris, wet or slippery spots, holes in the floor or stairs, nails, ash pits, and other hazards. Look up, down, and all around for hazards. There may be holes in the floor, loose boards, falling drywall, caving ceilings, swinging wires, or other hazards.

- Be alert to the possibility of gas leaks and hanging or swinging electrical wires or pipes. Even if you think the area has been ventilated and the gas turned off, it is not a good idea to use matches or other open flames, or anything that might generate a spark.

- The fire will have displaced many animals, and some may have taken up residence in your house. Watch for snakes and spiders (and other poisonous creatures, depending on where you live). Poke into possible hiding places with a stick. Remember that some very large creatures, such as bears, may have entered looking for food and shelter.

- Of course, if you smell gas, leave the building at once. Don't switch *anything* on or off, including phones, power, lights, water, or other potential spark-generating devices. Don't even open a window to try to ventilate the gas smell; this could cause a static discharge and ignite the gas. Call your gas or propane supplier at once *after* you have left the area.

- For the same reason, it's not a good idea to use the building's phone or your cell phone, PDA, or pager. These are sources of sparks that could ignite lingering gas fumes.

- It's a good idea to check the entire house and walk around the outside perimeter regularly to make sure hidden sparks have not re-ignited. Burning embers can blow in from quite a distance. Remember underground tree roots may also be burning or smoldering long after a fire appears to be out.

- You also need to get out of the house from time to time to get some fresh air. If any smoke or other fumes are present inside the house, you may become disoriented and not realize the air is foul inside. Getting outside for at least 15 minutes every hour and away from any lingering smoke is a good rule of thumb.

- Contact your utility company, power cooperative, propane supplier, telephone company, or other utility providers immediately if you spot or suspect a problem such as swinging wires, broken pipes, or the smell of gas. **Note:** Your power or water may already be off because the firefighters or other authorities threw the master breaker or because of downed or damaged power lines or broken pipes. When you have verified it is safe to turn the electrical power on again, if it does not work, the main breaker may have tripped. Fires have a tendency to cause this. You should have an electrician or mechanic check out all appliances or anything with a motor before you try to turn anything on, flip any light switches, or plug anything into an electrical outlet.

You should already have contacted your insurance company when you were ordered to evacuate. Even if you haven't, remember that most of the clean up work (debris removal, clean up, repairs, and replacement, as well as safety checks, such as the electrical system) should be covered by your business or homeowner's insurance policy.

Checking Outside Around Your Structures

Trees

When you walk the perimeter of your house, you might want to extend the inspection so you carefully check every tree on the property. (Be especially careful to avoid stepping into an ash pit, the burned out hole where a tree once stood that is now covered over with hot ashes.) You may want to hire a tree service if you have lots of trees to check, or need to remove trees. Trees that appear to be extensively damaged may actually survive the effects of a wildfire. Others that appear relatively okay may actually have damage you can't see. In general, tree survivability is determined by the amount of damage to the crown, bole, and roots. The best sources of information and assistance to help you assess the damage are:

- Your county extension office or land-grant university.
- Your state forestry agency (See contact information list beginning at page 157).
- The USDA Forest Service's Forest Health Protection Program (See http://www.fs.fed.us/foresthealth/.)
- A well-established and reputable local tree service or landscape specialist.

Trees, grasses, and other plants are essential to maintain the stability of the ground. They help prevent erosion and minimize the likelihood of, and damage done by, floods and mudslides. So, it's advisable to have professionals make the decision about what to keep and what to remove from your landscaping. You may think something is destroyed when it isn't, or you may not know what the most suitable choices are to enhance your property or protect it from erosion. However, if you'd like to begin by checking out your trees yourself, here is a list of a few of the things you should look for:

- **Obvious burns on the trunk or branches.** If there are burns around the entire trunk, or deep burn holes in the trunk, the tree probably will not make it. Mark it with colored tape for later removal.
- **Burned roots.** Use a shovel or metal rod and carefully probe the root area about 6 to 8 inches down to see if the roots have burned. If the roots have burned, mark the tree for later removal. Some fires burn underground for months or even years, so it is wise to be cautious. If you are not comfortable checking for burned roots, call a tree service or your state or country forester for advice. If the ground has crystallized or is too hard to probe, call a tree service to do the job. And save your receipts—your insurance company may cover some of the costs of the tree service.
- **Scorched trees.** Most trees will recover if they are only partially burned or seared from heat. But it usually takes an expert, such as a representative of the USDA Forest Service, your state forester, or a tree service company to determine whether a tree will survive. The following chart summarizes and compares survivable damage for Ponderosa Pine and

Douglas Fir trees.[27] But this is only a sample; you probably have a much broader mix of trees in your area. Check your state department of natural resources, the USDA Forest Service, or your state forester for information specific to your region.

Type of Tree/ Damage	Ponderosa Pine	Douglas Fir
Crown Scorch	Ponderosa pines may survive up to 75% crown scorch if fire occurs later in the summer, after buds have set for the following year. Long needles provide protection for developing buds.	Douglas-fir with more than 50% crown scorch, particularly if developing buds have been destroyed, are less likely to survive.
Bole Char	Ponderosa pine bark is less readily damaged by fire; but damage depends on size and vigor of tree. If inner bark is destroyed on more than 50% of bole circumference, survival is unlikely.	Douglas-fir bark, on mature trees, is less readily damaged by fire; but damage depends on size and vigor of tree. If inner bark is destroyed on more than 50% of bole circumference, survival is unlikely.
Root Damage	Damage to roots or root collar, to the extent that inner bark (cambium) is destroyed on more than half of tree's circumference or half of major lateral roots, will usually result in tree's death.	Damage to roots or root collar, to the extent that inner bark (cambium) is destroyed on more than half of tree's circumference or half of major lateral roots, will usually result in tree's death.

Water

If your home and property were not damaged by fire, your water may be safe, but it is better to assume otherwise until you have it checked. In fact, it is probably a good idea to presume your water source is contaminated, particularly if you depend on a well (your own or a community well) until told otherwise. If you are on a well, you should arrange for a private service to test the water. In some cases your county may provide this service, but it can take quite a long time and you'll need to boil or haul water in the meantime.

If you are on a public water system, you should contact your county health office for advice. Your sewer or septic tank may have backed up into the water line if water pressure was lost, causing bacteria to enter the water

---

[27] Data compiled from information presented on the USDA Forest Service website pertaining to the Hayman Fire at http://www.fs.fed.us/r2/psicc/hayres/index.htm under the Ponderosa Pine and Douglas Fir Survivability links.

pipes. Ground water may have seeped into broken pipes, or other causes may have affected your water supply. Always assume the water is not safe until you can find the appropriate civil authority to tell you otherwise, or you have the water tested privately.

Your county health department should be able to suggest water testing services, or you can look in the phone book. Referrals from neighbors are also good choices, or your insurance company may have recommendations. Again, as with any other contractor, it is probably a good idea to check out water testing services with your local Better Business Bureau.

YOUR RIGHTS AND RESPONSIBILITIES AFTER A LOSS

Losing your home and your personal property in a fire can be devastating. What would be worse is not having adequate insurance to cover your losses. See Chapter 5: Insurance for a complete discussion of the types of policies and coverages available. But even with adequate coverage, you may simply make your problems worse if you fail to understand your policy and take the appropriate steps at the proper time to enforce your rights.

Generally, there are three considerations you should keep in mind when dealing with property losses and insurance claims:

- **Communicate.** Let your insurance company know as soon as possible after your loss of your potential claim and collect all the information you need from them to file the claim, including deadlines and the necessary forms.

- **Corroborate.** Back up your claim with documentation, such as original purchase receipts, "Before" and "After" photographs or videos, and subsequent repair bills.

- **Cooperate.** Follow the insurance company's guidelines and procedures for filing a claim and work with company representatives to complete the needed repairs, replacement, and rebuilding.

As suggested above, don't wait to find out whether or how badly your property has been damaged to call your insurance agent. *(Communicate.)* One very good reason for this is that your homeowner's insurance probably covers the cost of meals and lodging (i.e., additional living expenses) when you are forced out of your home due to a fire. So call your agent as soon as you are evacuated and safely out of the fire danger area—and save your hotel and food receipts.

It's also a good idea to keep receipts for *all* your expenses—even clothes and toys. *(Corroborate.)* Many expenses, such as those you incur to protect your property from additional damage, may very well be recoverable as part of your property or auto insurance claim. Without receipts, however, these expenses will probably *not* be covered.

It's up to you to understand your insurance policy and what it provides. Your policy imposes timelines you must follow to receive maximum coverage and repayment under your insurance policy. And you must take a number of steps, often in a particular order and by certain deadlines, to ensure your covered losses are actually paid by your insurer. *(Cooperate.)* Some of these steps include:

- Give written notice to your insurance company or agent about the loss, generally within 30 to 60 days. This includes copies of evacuation orders if you plan to use the additional living expenses or loss of rents feature of your policy.

- Arrange to meet with your insurance company adjuster at the site of the loss.

- Protect your property from further damage. If applicable, make emergency repairs and keep your receipts to add to your claim. If necessary, turn off utilities (such as gas or propane) before you leave to prevent collateral damage. At the same time, you must do only what is reasonable and follow the instructions of law enforcement and fire authorities. For example, if you turn off the gas, and thus the heat, during cold weather and your pipes freeze and burst, your loss may not be covered. However, if you were ordered to do so by the fire authorities and can prove it, your loss probably will be covered, even if the fire never touched your structures or premises.

- Make a list of all damaged or destroyed property, including as much detail as possible about each item, such as the original price, age and condition at the time of the fire, and the extent to which it was damaged or destroyed. This is where your "Before" photos or videos and the receipts you copied will come in handy.

- Comply with all insurance company representative requests.

Your insurance company may (but does not have to) assist you with finding contractors to: clear away debris; perform any needed demolition; remove and store undamaged contents; clean and restore repairable items; reconstruct buildings, fences, garages, and other structures; and perform other necessary actions. Generally, *you* have to contract with these companies directly, but your insurance company will often give you a list of "preferred" vendors with which the company has worked in the past. The advantages to using one of these vendors are (1) the vendor is familiar with how your insurance company wants things done, and (2) the vendor will generally agree to wait for payment until you are paid by the insurance company, provided the insurance company issues you a two-party check (one that is payable to you *and* the vendor).

Clean-up and Demolition

If firefighters have to douse your house or punch holes in the walls to save the structure from fire, you will be amazed at how much cleanup and repair work will need to be done, beginning with demolition of the damaged portions. This can be the hardest part of the job and one of the messiest. Drywall gets dragged across the deck, carpeting sits in soggy mounds around the house, and fixtures get knocked down and broken. A "cleanup" crew might toss out anything that isn't bolted down, and even some things that are—towel racks, shower rods, area rugs—even though these items were salvageable. So it's important to remove anything from the house that isn't damaged or appears to be salvageable. You wouldn't want the contractor to inadvertently damage or discard something you want to keep.

You should remove all burned items from the house after the adjuster says it's okay and you've photographed or video taped the items for your claim.

It is a good idea to move everything that is unburned or salvageable to a self-storage facility, another building on your premises, or a cleared space (and cover it with a tarp) far enough away from the damaged structure so that these items are not in the way of the demolition crew. Of course, keep your receipts for the storage facility and the moving costs for insurance purposes. It is probably a good idea to take any valuable items to a secure location (perhaps your temporary living quarters).

Once the work is underway, you may find it good idea to *stay out of the way and don't micro-manage.* You will get far better cooperation if you tell the contractors in advance when you plan to visit the work site so you interfere with their work as little as possible. If you are cooperative, your job will probably get done faster and better than if you stay underfoot and question everything. You may find that this courtesy is repaid with a few unexpected services (such as moving a light switch to a more appropriate location—easy to do when new drywall is going in).

If you do any of the demolition and clean-up work yourself, be sure to wear rubber gloves to protect your hands from toxins and other chemical irritants. Good lighting is helpful for the cleanup process; unfortunately, you may not have electricity. If you bring a generator onto the premises, make sure there are no gas leaks on the premises and avoid using it (or anything else that runs on gas, propane, or diesel) inside a structure.

Note: If you don't have insurance and have to do the demolition, clean up, removal of property, and storage yourself, or you pay your own temporary living costs, keep all your receipts and keep track of your mileage. Some of these costs may be partially tax-deductible, but you'll need excellent records to claim the deduction.

Choosing the Right Contractor

Having your home, office, or property damaged by a wildfire is frustrating, but sometimes getting the repairs done in a timely manner by reputable companies can be even harder on your mental state and patience. Unfortunately, disasters can bring out the very worst in some people, leading them to seek out people who have already suffered a loss with the intent of making matters worse in search of a quick buck.

What can you do to protect yourself, and what should you watch out for?

If you are working with or through your insurance company, the insurance adjuster or someone at the main office may be able to direct you to reputable businesses. Some insurance companies have a "preferred vendor" program whereby the insurance company suggests a contractor (or multiple contractors). If you follow their recommendation, they will pay the contractor directly, making your life much easier.

The fact that a vendor has done lots of work for the insurance company and made the company's recommended list does not guarantee the performance of the contractor or mean they will always do quality work. When disaster recovery operations get underway, even the best contractors have to hire additional people, some perhaps with unclear references. However, if you have the insurance company and the general contractor on your side, the costs associated with any substandard work can usually be charged to the responsible party. Most importantly, you may not have to pay first and fight for a refund later.

Unfortunately, even if you find your own contractors, you *may still* become a target. Whether you also become a victim is largely up to you. Here are some tips that may keep you from becoming a victim:

- *Always* ask your colleagues, friends, neighbors, and insurance company for *referrals*.

- *Always* ask for, and *check,* references. Try to visit finished job sites and work-in-progress sites.
- Check out each contractor with the Better Business Bureau (BBB).
- Insist on a written estimate.
- Insist on a written delivery or completion schedule.
- Don't put down more than 15 to 20 percent of the written estimate amount. That is the most a reputable contractor should charge.
- Ask how your project will be supervised. Make sure you meet the supervisor, and get his or her direct contact information (preferably e-mail and cell phone).
- Use the phone book only as a last resort, ask for references, and check with the BBB.

How do you know if a contractor will be reputable, reliable, and honest? Unfortunately, you don't; but following these steps will improve your odds.

RELATIVELY UNDAMAGED PROPERTY

You are probably one of the lucky majority whose property survives a fire with little or no *visible* damage. The land around your buildings may have been damaged, but the firefighters were able to save your buildings. However, there are still safety issues you need to consider and *you should take the same precautions* listed earlier in this chapter when entering your property. Exercise extreme care around any structure in a fire-damaged area. You may want to consider having a building inspector verify the condition and safety of any structures.

You may also have some water or smoke damage to your home and other structures, as well as damage to some of your personal belongings or vehicles. If you find that law enforcement or fire authorities have turned off your utilities, ask if it is safe to turn them on again. Be very careful using open fire or devices that may generate sparks (cell phones, chain saws, light switches, generators, etc.). Be careful when you turn on any disconnected utilities or flip circuit breakers. Additionally, you may want to have your water supply checked to be sure it is safe.

If you were forced to evacuate, you should already have opened an insurance claim file in order to recover your temporary lodging costs. If not, you still need to notify your insurance company that your property was involved in a fire in case you later discover damages to your belongings, structures, or vehicles. It's a good idea to save any receipts for costs you incur; some expenses may be recoverable. For example, your computer may not be working properly, but because you didn't use it right away it isn't on your initial claim. After you take it to the repair shop and discover that the problem was related to the fire (including damage from water, smoke, or firefighting efforts), you should be able to claim the repair expense.

PREPARING FOR FLOODS, MUDFLOWS, AND LANDSLIDES

Keep in mind that floods, mudflows, and landslides sometimes follow in the aftermath of a wildfire. So putting out the fire may not be the end of your woes. If the fire passed through your neighborhood, the ground has probably been denuded of trees and plants that used to hold back the water and stabilize the soil. The scorched earth may even have crystallized if the fire was hot enough, so water no longer soaks in but runs right over it. This multiplies the typical runoff many times over, creating a flood. Most floods that occur following a wildfire are of the flash-flood variety and are incredibly destructive.

Areas in Colorado devastated by the Hayman Fire in 2002 and the Buffalo Creek Fire in 1996 were further damaged by a series of floods in the following months. Some residents then discovered that homeowner's insurance does *not* cover floods. So you may want to consider purchasing flood insurance and, possibly, earth movement insurance, if you have not already done so. (See the information about flood insurance in Chapter 5: Insurance.) Note that you must purchase flood insurance at least 30 days before a flood to be covered for damages.

Basic Steps to Minimize the Risk of Flood Damage

Thoughtful building and landscaping with an eye toward possible floods are important factors in protecting your property. Here is a quick checklist of things you can do to minimize the risk of flood damage to your home. If your house is already built, some of these steps may not be feasible, but keep them in mind if you have to rebuild.

- When building new structures, position them away from low spots and gullies on your property where runoff is likely to collect.
- Plan for adequate drainage by contouring the ground to slope away from your house and other structures.

- Plant foliage and ground cover that will develop good root systems, so even if they are burned, they will help hold the soil together and help it absorb water. Check with your county agricultural office, state forester, or local botanical garden for recommendations on the types of plants to use.

- As a temporary measure, while you're waiting for new plants to mature, install sandbags or build earthen diversion berms around your home and other structures.

- Build a water diversion system with drainage channels, tiles, and retaining walls. A civil engineer or landscape designer can help plan your system.

- Position your propane tank at a lower level on your property than your house is situated so it does not break free in a flood and roll into the house. During the spring runoff one year, my propane tank came loose and rolled almost into my kitchen window.

- Clear debris from your property—it can be picked up by the floodwaters and move with incredible speed, punching holes in your structures.

- Store valuable personal belongings upstairs, not in your basement. Property in basements is almost never covered, even if you have flood insurance.

- Ask your insurance company for practical suggestions.

Basic Steps to Prepare for Mudflows and Landslides

Mudflows occur when the soil retains too much water, it becomes unstable, and starts to move downhill. Landslides occur when slopes, eroded by water, give way, and rocks, dirt, debris, and trees slide downhill. So take these facts into consideration when planning your landscaping. Try to achieve a balance between the water that is retained on your property and the water that is diverted. A civil engineer can help you make this determination and plan accordingly.

Mudflows and landslides can be even more destructive than floods because of the amount of dirt and debris they carry with them. The damage they cause can include:

- Breaking or exposing electrical, water, sewer, gas, and sprinkler lines.
- Upending fuel storage and water tanks.
- Cracking or disrupting roads and rail lines.
- Washing away bridge supports.

- Clogging culverts and filling drainage ditches.

- Piling debris around your property or crashing it into your house.

- Carrying contaminated water and sewage into dwellings and water lines.

Interesting Note

The most expensive landslide in U.S. history happened in Thistle, Utah in the spring of 1983. It was caused by groundwater buildup from heavy rains the previous September and the melting of deep snowpack from the winter of 1982-83. Within a few weeks, this slow-moving landslide dammed the Spanish Fork River, obliterating U.S. Highway 6 and the main line of the Denver and Rio Grande Western Railroad. The town of Thistle was inundated under the floodwaters rising behind the landslide dam. The landslide was as deep as 1/2-mile and ranged from 1,000 feet to almost one mile in width. The estimated cost to repair the damage to roads, bridges, and other structures was over $500 million.[28]

ADDITIONAL ASSISTANCE

If your property was damaged or destroyed by fire or flood, and you have no structural or flood insurance, or you have exhausted your insurance coverage, there are several places you can turn to for assistance with rebuilding, replacing your personal belongings and business property, obtaining mortgage or rent assistance, and getting rebuilding loans. These include:

- The Federal Emergency Management Agency (FEMA).

- State emergency relief agencies.

- The Small Business Administration (SBA).

- The American Red Cross (ARC).

- The Salvation Army.

- The Natural Resources Conservation Service (part of the U.S. Department of Agriculture).

- Presidentially declared disaster area programs.

---

[28] Information from the FEMA website at
http://www.fema.gov/hazards/landslides/landsli.shtm and a USGS press release at
http://www.usgs.gov/public/press/public_affairs/press_releases/pr1133m.html.

FEMA and State Agencies

The Federal Emergency Management Agency (FEMA) and your state's office of emergency management are valuable resources for wildfire recovery links and information. The FEMA website is located at http://www.fema.gov/. Your state site should be accessible by doing an Internet search in *Google* for [state name + emergency + assistance + agency] or something similar. Some of the services FEMA provides are:

- Flood insurance through the National Flood Insurance Program (NFIP) (http://www.fema.gov/nfip/).

- Individual and Family Grants—assistance to cover a variety of disaster-related expenses (those not covered by insurance or for the uninsured/underinsured) such as:

  o Certain home repairs.

  o Certain medical personal property such as wheelchairs.

  o Replacement or repair of certain personal property.

  o Essential tools required for work.

  o Medical, dental, or funeral expenses.

  o Transportation.

According to FEMA,[29] the IFG program is intended for residents whose needs exceed the help available through insurance, government and volunteer programs. The IFG program is part of the overall disaster assistance effort triggered by a presidential disaster declaration. It was created to provide people, impacted by disasters, the necessary expenses and serious needs that cannot be met through insurance or other forms of disaster assistance, including low-interest loans from the U.S. Small Business Administration.

If you are in a presidentially declared disaster area, there may be tax breaks and filing deadline extensions, as well as an opportunity to amend the previous year's tax return to take advantage of special breaks for certain tax deductions. If you think you may qualify, check with your tax advisor and your state emergency management office.

The maximum allowable grant is $25,000 per applicant. FEMA provides 75 percent of the grant money, and your state provides the other 25 percent. These IFG payments are usually only available to people who are

---

[29] See http://www.fema.gov/rrr/inassist.shtm and FEMA-DR-1421-PR18, dated July 9, 2002, at http://www.fema.gov/diz02/d1421n18.shtm.

not eligible for or have been denied low-interest Small Business Administration disaster loans.

- Disaster Unemployment Assistance through your state Division of Labor if you lost income because of the wildfires. You must be ineligible for regular unemployment programs. There are application deadlines associated with each disaster, so it is important to visit the FEMA website to get the necessary information and procedures to apply. This assistance is particularly useful for sole proprietors and many independent contractors. Note that if you should have been paying unemployment insurance premiums and chose not to, you are probably not eligible for this assistance.
- Fire Management Assistance declarations to help state and local governments and Indian tribes minimize effects and control/manage fires. Check the FEMA website for dollar limitations.

Insurance Coverages—Auto and Business

Some damage is not covered under your homeowner's, renter's or condominium insurance policy, or your fire policy (for non-owner occupied structures). In other cases, the coverage may involve *both* your homeowner's policy and your business or auto policy. As with any insurance coverage and policy requirements, specifics vary both by state and between insurance companies.

Your auto policy should cover damage to your personal vehicles. This normally includes water, smoke, heat, fire, looters, or other damage directly related to the fire or the results of firefighting. However, you *must* have "other than collision" coverage, sometimes referred to as "comprehensive" coverage. If personal or business property inside your vehicles is damaged or destroyed, the cost almost always fall under your homeowner's policy or business insurance policy. It is a good idea to notify your insurance company as soon as you discover or suspect your vehicle (or its contents), trailers, or similar items have been stolen, damaged, or destroyed.

If you do not have homeowner's insurance but you do have a business insurance policy covering fire and other hazards, you should be able to file a claim for the damaged or destroyed items used in your business against that policy. Note that homeowner's policies may provide some coverage for business property and files, but this coverage is extremely limited. One item that may be covered by both a business and a personal policy is your personal computer, even if it is used for business. But there are limits, usually $3,000-5,000. Notify your insurance company as soon as you discover or suspect your business property has been stolen, damaged, or destroyed.

Small Business Administration

The SBA is good source of low-interest rate disaster loans, but not all victims are eligible. To find out if you qualify, visit the SBA Disaster Assistance page at http://www.sba.gov/disaster/. The SBA website, located at http://www.sba.gov/, also has a lot of useful information in the form of answers to frequently asked questions (FAQ) and is worth a visit. You can also contact the SBA at 1-800-U-ASK-SBA (1-800-827-5722); Fax: 202-481-6190; TDD: 704-344-6640; E-mail: answerdesk@sba.gov.

If you do not qualify or are turned down, you should approach FEMA about the Individual and Family Grant program discussed in the "FEMA" section above.

According to the SBA website:[30]

Personal Loans

As an individual, there is one basic loan, with two purposes, available to you:

- Personal Property Loan: This loan can provide a homeowner or renter with up to $40,000 to help repair or replace personal property, such as clothing, furniture, automobiles, etc., lost in the disaster. As a rule of thumb, personal property is anything that is not considered real estate or a part of the actual structure. This loan may not be used to replace extraordinarily expensive or irreplaceable items, such as antiques, collections, pleasure boats, recreational vehicles, fur coats, etc.

- Real Property Loan: A homeowner may apply for a loan of up to $200,000 to repair or restore their primary home to its pre-disaster condition. The loan may not be used to upgrade the home or make additions to it. If, however, city or county building codes require structural improvements, the loan may be used to meet these requirements. Loans may be increased by as much as 20 percent to protect the damaged real property from possible future disasters of the same kind. Note: A renter may apply only for a personal property loan.

---

[30] See http://www.sba.gov/gopher/Disaster/homeall.txt and http://www.sba.gov/gopher/Disaster/pdball.txt.

Insurance Proceeds: If you have insurance coverage on your personal property/home, the amount you will receive from the insurance company will be deducted from the total damage to your property in order to determine the amount for which you are eligible to apply to the SBA. If you are required to apply insurance proceeds against an outstanding mortgage, the amount applied can be included in your disaster loan. If, however, you voluntarily apply insurance proceeds against an outstanding mortgage, the amount applied cannot be included in your disaster loan. If you have not made a settlement or are having trouble reaching an agreement with your insurance company, you may apply for a loan in the full amount of your damages and assign any insurance proceeds to be received to the SBA.

Interest Rates on Loans: The law requires a test of your ability to obtain funds elsewhere in order to determine the rate of interest that will be charged on your loan. This credit-elsewhere test also applies to applicants for both personal property and real property loans.

Applicants Who Can Obtain Credit Elsewhere: The interest rate to be charged is based on the cost of money to the U.S. government, but will not be more than 8 percent per year.

Applicants Determined Unable to Obtain Credit Elsewhere: The interest rate to be charged will be half of the interest rate charged to applicants determined to be able to obtain credit elsewhere, but will not be more than 4 percent per year.

Term of Loan: The maximum maturity, or repayment term of an SBA loan, is set at 30 years. However, the SBA will determine repayment terms on a case-by-case basis according to your ability to repay.

Business Loans

If your business—large or small—has suffered physical damage as a result of a disaster, you may be eligible for financial assistance from the U.S. Small Business Administration. Any business that is located in a declared disaster area and has incurred damage during the disaster may apply for a loan to help repair or replace damaged property to its pre-disaster condition. The SBA makes physical disaster loans of up to $1.5 million to qualified businesses.

Use Of Proceeds: Repair or replacement of real property, machinery, equipment, fixtures, inventory and leasehold improvements may be included in the loan. In addition, disaster loans to repair or replace real property or leasehold improvements may be increased by as much as 20 percent to protect the damaged real property against possible future disasters of the same type. SBA loans will cover uninsured physical damage. If you are required to apply insurance proceeds to an outstanding mortgage on the damaged property, you can include the amount applied in your disaster loan.

Interest Rates: The interest rate that the SBA charges on a disaster loan is determined by your ability to obtain credit elsewhere—that is, from nonfederal sources.

If You Cannot Obtain Credit Elsewhere: If the SBA determines that the business (or nonprofit organization) is unable to obtain credit elsewhere (considering the cash flow and assets of the business, its principals and affiliates), the law sets a maximum interest rate of 4 percent per year. The maximum maturity for such business disaster loans is 30 years. However, the actual maturity is based on your ability to repay the loan.

If You Can Obtain Credit Elsewhere: For businesses that the SBA has determined are able to obtain credit elsewhere, the interest rate cannot exceed that being charged in the private market at the time of the physical disaster or 8 percent, whichever is less. The maturity of this loan cannot exceed three years.

Note: Charitable, religious, nonprofit and similar organizations with the ability to obtain credit elsewhere are eligible for physical disaster loans for up to 30 years at an interest rate based upon a different statutory formula. The nearest SBA disaster office can supply you with the interest rate.

The American Red Cross

The American Red Cross (ARC) provides vouchers, redeemable with *any* merchant, at no cost to disaster victims. You can check the ARC website, http://www.redcross.org/, for information about getting vouchers. These vouchers are paid for with donations collected throughout the year from individuals, businesses, and other groups. According to its website, the Red Cross does not solicit or accept small-scale donations of supplies or other goods. Large-scale donations are sometimes sought, however, and monetary donations are always needed.

The Red Cross does not generally give out supplies or other goods; its philosophy is that people need to reassert control over their lives as quickly as possible, and vouchers allow them to make their own choices. Vouchers also return money to the local economy to help the community recover from a disaster, too. To find out more about what the Red Cross offers and how you can help, visit their website. Keep in mind that another major service provided by the Red Cross is coordinating and managing blood donations and supplying medical facilities around the country with donated blood and blood products. As the events of 9/11 showed, the need for blood is always present, and donating blood is one way you can help others recover from wildfires and other disasters or emergencies.

All American Red Cross disaster assistance is free, made possible by voluntary donations of time and money from the American people. The Red Cross supplies nearly half of the nation's lifesaving blood, also made possible by generous voluntary donations. You can help the victims of thousands of disasters across the country each year by making a financial gift to the American Red Cross Disaster Relief Fund, which enables the Red Cross to provide shelter, food, counseling, and other assistance to those in need. You can make a secure online credit card donation by going to the ARC website or call 1-800-HELP NOW (1-800-435-7669) or 1-800-257-7575 (Spanish). Or you may send your donation to your local Red Cross or to the American Red Cross, P.O. Box 37243, Washington, D.C. 20013. To donate blood, please call 1-800-GIVE-LIFE (1-800-448-3543), or contact your local Red Cross to find out about upcoming blood drives.

The National Resources Conservation Service

This is an arm of the U.S. Department of Agriculture. Their website is http://www.nrcs.usda.gov/. This agency helps you minimize collateral damage to your property from floods, mudflows, and landslides, as well as offering assistance in the preparation aspects of wildfire mitigation.

The Salvation Army

The Salvation Army has been assisting individuals, businesses, and communities with disaster recovery services since at least 1900, when it stepped in to help victims of the disastrous Galveston hurricane. Some the services available through the Salvation Army (http://www.salvationarmyusa.org/) include:

- Counseling victims, consoling the injured and distressed, comforting the bereaved. (Don't overlook the importance of counseling in dealing with the emotional stress you're likely to suffer.)

- Acting as an information clearinghouse to find assistance, locate displaced people, and so forth.
- Providing mobile "feeding stations" to prepare and serve hot meals to victims and volunteers.
- Providing financial assistance in the form of grants to victims who demonstrate need.
- Establishing and maintaining shelters at appropriate locations to house evacuated persons.
- Offering programs such as free childcare to enable victims to return to their property and begin salvage operations, apply for government (FEMA, SBA) and other assistance programs, schedule rebuilding (helping victims deal with contractors), and perform other tasks to return to a normal lifestyle.
- Soliciting and distributing water, nonperishable foods, furniture, housekeeping supplies, building materials, and so forth.
- Assisting, in some cases, with reconstruction efforts.

Unlike the American Red Cross, the Salvation Army actively solicits and accepts donations of basic supplies from individuals.

PRESIDENTIALLY DECLARED DISASTER AREA PROGRAMS

This is a relatively new tax relief program passed in response to the severity of the 2002 wildfire season.

What is a Presidentially Declared Disaster Area?

A presidentially declared disaster is a disaster that occurs in an area declared by the President to be eligible for federal assistance under the Disaster Relief and Emergency Assistance Act.

When to Deduct the Loss From a Presidentially Declared Disaster

If you have a casualty (property/fire/flood, etc.) loss from a disaster that occurred in a presidentially declared disaster area, you can choose to deduct that loss on your return or amended return for the tax year immediately preceding the tax year in which the disaster happened. If you make this choice, the loss is treated as having occurred in the preceding year. This will probably result in a significant tax refund. You will have to file an amended return. For example, if you were victim of the 2002 wildfires, and you are in a presidentially declared disaster area, you could have amended your 2001 tax return for a refund, or you can declare the loss on your 2002 return. Always consult a tax advisor before making these decisions. Tax preparation software may or may not pick up this benefit.

The IRS may postpone for up to 120 days certain tax deadlines of taxpayers who are affected by a presidentially declared disaster. The tax deadlines the IRS may postpone include those for filing income and employment tax returns, paying income and employment taxes, and making contributions to a traditional IRA or Roth IRA.

If any tax deadline is postponed, the IRS will publicize the postponement in your area and publish a news release, revenue ruling, revenue procedure, notice, announcement, or other guidance in an Internal Revenue Bulletin.

See IRS Publication 547 for more information. You can order it through http://www.irs.gov/. There are deadlines for filing, so you need to read the publication carefully or let your tax professional advise you.

DEALING WITH THE EMOTIONAL AFTERMATH

The first part of this chapter deals extensively with the physical reconstruction and recovery needs you may face. For many people, there are additional mental and emotional needs to be addressed. There are a number of places to go for assistance and counseling, including:

- Your health insurance provider.
- The Salvation Army.
- Community outreach programs, such as the Gathering Place and the Mental Health Sanctuary (see the following paragraph).
- Your church or other religious organization or counselor.
- Other providers, located through your state, county, or city community services office.

A useful reference website is http://www.mhsanctuary.com/mh/toll.htm. This site, called the Mental Health Sanctuary, has a comprehensive list of toll-free resources and hotlines. Some of the more relevant listings include an Attorney Referral Network, Consumer Credit Counseling Services, an "Emotional Distress" hotline, Housing and Urban Development (HUD) hotline, and the National Mental Health Association. The list on this website is quite comprehensive and useful.

CONCLUSION

Returning to your property after a wildfire can be both difficult emotionally and unsafe. Hopefully, the suggestions in this chapter have given you enough information to guide you through this phase with care, get the help you need to rebuild, and to find additional resources for recovery.

Chapter 5: Insurance

PROPERTY INSURANCE

A typical property insurance policy covers, or may be written to cover, a variety of losses. These losses may include damage to your home from fire, wind, or water; theft or loss of your personal belongings; your liability if someone is injured on your property; and even additional living expenses to pay for a hotel in the event your house is rendered uninhabitable by a covered loss or by a government-ordered emergency evacuation. Fire insurance is just one component of your homeowner's, renter's, condominium, or business insurance. But if you live in the wildland/urban interface, fire insurance will be the most important component of your policy when wildfires threaten.

This chapter explains some of the more typical policy provisions and includes a chart of commonly included and excluded coverages. It is important to note that insurance is primarily regulated by state governments, so there will almost certainly be some differences between the generic information presented here and the exact policy provisions and requirements where you live. In addition, there may even be significant differences among the various policies of a single insurance company and from one company to the next. Your insurance advisor is the best source of specific insurance policy information.

Who Should Have Property Insurance?

If you own property in the wildland/urban interface—or anywhere else for that matter—*you* need property insurance. Without it, you are not covered for property losses due to fire. Whether you own a house, townhouse, condominium, business property, or merely a lot with a building or two on it, you should have property insurance. Even if you don't own real property and are renting your home, you should have a renter's policy that covers the loss of your personal belongings due to fire. You may think you have nothing of value, but a quick trip through your local shopping center, with an eye to replacing your clothes, electronics, furniture, books, kitchen stuff, appliances, and so forth, should be a real eye-opener.

Two of the saddest situations you'll hear about following a wildfire are:

- The homeowner who did not have property insurance because her house was paid for and there was no longer a mortgage company requiring the insurance.

- The tenant who did not have renter's insurance because he did not know about it, chose not to buy it, or assumed that the landlord's policy covered everything.

These types of tragedies are reported in the news after every major fire, and still people choose to go without proper insurance. But the coverages are so comprehensive and the costs are so reasonable, it just does not make sense to drop, or elect not to have, a property insurance policy that includes fire coverage.

What Happens If I Don't Have Property Insurance?

There will always be people who choose to go without insurance of one kind or another. In the insurance industry, this is called "going bare" or "going naked." If you drive a car, you have no choice. Every state requires, at a minimum, that you have basic liability coverage, even if your car is paid for. However, once you pay off your home mortgage, there is typically nothing that requires you have property insurance. And if you're a renter, it's rare that you would be required to have renter's insurance (although some residential leases do require it).

If you do not have property insurance and a disaster, such as a wildfire, befalls you, there are government agencies and charitable organizations that may be able to offer limited assistance with rebuilding, refurnishing, and temporary living expenses. These are discussed at some length in Chapter 4: Recovery. And although many people—such as your neighbors, friends, and family—will try to help, their resources may be limited and will never come close to the support and dollars available through a reputable insurance company.

Keep in mind that, without property insurance, the burden of such tasks as clearing debris, finding contractors, supervising reconstruction, and arranging payments will fall entirely on you. When a water line break caused extensive damage to my home several years ago, I found out just how much a good insurance company can do. I don't know how I would have handled the cleanup and repairs, dealing with numerous contractors, or even making my own living arrangements without the help of my insurance company and its adjuster. I also would have hated to mooch off of my friends for the 12 weeks it took to fix everything!

Where Can I Get Property Insurance (Including Fire Insurance)?

Availability of insurance has recently become an issue in many states that have suffered from wildfires. People living in wildland/urban interface zones have suddenly realized they need fire coverage. At the same time, insurance companies are understandably reluctant to issue policies while the fires are burning. This has even resulted in delayed home sales in some cases because the buyers have been unable to get insurance.

Most people get their homeowner's, condominium, renter's, business, or other property insurance through the agent who wrote their automobile insurance. While there is no requirement that you purchase all of your insurance policies through one agent or company, doing so can make record keeping, claims processing, and getting your questions answered much simpler. It may also save you some money since you may qualify for a multiple-policy discount. But if you don't already have a company or an agent, you can even shop for insurance on the Internet. The most important thing is that you *get the insurance—and don't let it lapse.*

Types of Coverage

Some of the key provisions that appear in a typical property insurance policy for an individual (as opposed to a business), including covered losses due to fire, address the following issues:

- Destruction / Damage to Dwellings
- Destruction / Damage to Separate Structures
- Destruction / Damage to Personal Property
- Loss of Use of the Property
- Additional Coverages
- Losses Not Insured
- Conditions—What is Required of You After a Loss

Dwellings

This type of coverage applies to dwellings and *attached* structures used as your primary residence. Materials and supplies on or adjacent to your premises are typically included if their intended use is construction or remodeling of your dwelling or related structures on your premises. This type of coverage does not include damage to or destruction of your automobile, trailer, camper, or similar items. Land or the value of land is also not covered. If the land must be restored in order to rebuild the structures, the cost related to the land is not covered.

Separate Structures

This is the coverage for detached garages, workshops, horse stables, and similar structures, provided they are not used for business purposes.

Personal Property

Coverage applies to personal property owned *or* used by the insured anywhere in the world. There are other options to cover the property of visitors to your premises. Note that this is how personal belongings in your car, trailer, or other vehicle are covered; they are not normally covered under your auto policy.

Special Limits on Certain Personal Property

This is the "fine print" part of your policy. Please read it carefully! You may have hundreds of thousands of dollars of coverage for your personal property (or that of other people in certain circumstances), but if the items have limited coverage or are excluded from coverage in your policy, you'll need a rider or additional coverage to insure their full value. Sample limits on coverage include:

- 10 percent of the Personal Property Coverage amount for items not at your premises when a loss occurs.
- $100 for cash or cash equivalents (including gold and silver coins).
- $1,000 for securities, passports, tickets, stamp collections, and similar items.
- $1,000 on watercraft and their associated items (you should have a separate policy for these, especially for liability purposes).
- $1,000 on trailers not used with watercraft.
- Jewelry, silverware, watches, furs, etc., need to be covered by a special endorsement or rider. Standard policy coverage limits are very low, especially for theft. If the loss is due to a fire, the items are normally covered under the available coverage for your personal property.
- Firearms need to be covered by a special endorsement or rider. Standard policy coverage limits are very low, especially for theft. If the loss is due to a fire, the items are normally covered under the available coverage for your personal property.

- Business property, other than electronic data processing equipment (computers, for example), losses are limited to $200. This is why you should have a separate business insurance policy, especially if you work from home. The total coverage for all electronic data processing equipment is usually limited to $5,000.

Loss of Use

This coverage is very important if you have to evacuate your home in advance of a fire, or if your home is damaged or destroyed by a fire and you have to live elsewhere until it's fixed. The three main components are:

- Additional living expenses (ALE). If a *covered* loss makes your premises unfit to live in, the cost to live in a temporary residence (such as a furnished apartment or extended-stay hotel) is covered for up to 12 months or the time it takes to repair *or* rebuild your primary residence, whichever is shorter.

- Loss of rents. If you are renting a place to someone else and a *covered* loss makes the premises unfit to live in, the policy will cover your lost rents. It does not cover your displaced renter's temporary lodging or property replacement costs. They will need their own renter's policy for that.

- Prohibited from use. If you are ordered to evacuate your home by civil authority because of possible damage from what would be a covered cause, the ALE provision of your policy should apply.

Additional Coverages

If, as a result of a covered loss, any of the following are required, the cost is normally covered:

- Debris removal.
- Emergency repairs.
- Replacement of trees, shrubs, plants, and lawns, except those used for business purposes (usually coverage for such items is limited).
- Fire department service charges.
- Emergency removal of property.
- Freezer food spoilage.

What Does Property Insurance Cost?

Costs vary widely from state to state, city to city, and even within the same ZIP code. The types of coverage you choose also affect the cost. For example, a full replacement provision or rider on your policy will probably increase your premiums by 10 to 20 percent per year, but the alternative is actual cash value (ACV). That means you get the current resale value of the item. For example, if your clothes are of good quality but several years old, and each pair of pants would cost about $60 to replace, you may find that the ACV is only $10 per pair. The benefits of additional coverages or riders are discussed in more detail later in this chapter.

If You Rent Your Living or Office Space

If you live in a rented house, apartment, or trailer, you could spend anywhere from $100 to $500 per year for a renter's policy that covers your personal belongings. In a few areas, the costs could be higher; but a good budgeting range is $200–300 per year. In addition to covering your belongings if they are damaged or destroyed, a renter's policy also covers theft (typically from your home or car) and may cover additional living expenses if you are displaced by a fire or other hazard. Your policy should also cover a certain amount of personal liability (perhaps $100,000) if, for example, someone trips and is injured in your kitchen or on your front steps. If you need additional personal liability coverage, there are "umbrella" policies that cover much more.

If you operate a business out of leased office space or a commercial building, your landlord will almost certainly require you to have a "business owner's policy," often referred to as a "BOP." These policies are fairly standard and normally cover repair or replacement of business property that is damaged or destroyed by fire or other hazards. They may also cover temporary relocation expenses if your business is displaced by a fire or other hazard, as well as some of the costs to reconstruct files and computer records. Note that some types of business equipment, records, and other items may be covered only by special endorsements or riders, so be sure to read your policy carefully. The landlord's property insurance policy should cover the structure itself, but it's a good idea to ask, before you sign the lease, whether your landlord's insurance coverage is adequate and up to date. Note that BOPs are intended for small businesses in selected industries and may cost as little as $250 per year. But they may also cost tens of thousands of dollars for some larger or riskier businesses. Giant corporations need far more elaborate (and expensive) policies.

Homeowner's Insurance

If you own a home and have a mortgage, you will be required by the lender either to provide proof of an adequate homeowner's insurance policy (naming the lender as the loss payee), or to buy insurance through your lender. If you obtain your own insurance, *you* are responsible for keeping a current certificate of coverage on file with the lender. Your insurance company may do this for you, but don't count on it.

If you have paid off your mortgage, you *can* drop your homeowner's insurance coverage, but the risks of incurring what would be a "covered loss" far outweigh the minor cost savings from having no insurance premiums to pay.

The cost of your policy is based on a number of factors:

- The cost to *rebuild* your home. This is *not* the same as the price your house would fetch if you were to sell it. The rebuilding cost is almost always less than the market value since the ground on which your house sits isn't destroyed, and certain parts of your house may survive a fire, such as a well, a septic field, or the foundation.

- The value of your personal belongings, and whether you elect to pay the extra cost for "full replacement" insurance.

- How far you live from a fire station and the nearest fire hydrant or similar large-volume water source. Many houses in the wildland/urban interface are downgraded because of their distance from a fire station or water source. This "deficiency" can double or triple your insurance costs.

- The past fire history of your neighborhood may also influence your property insurance costs. In addition, if you are in a "high crime" area (usually not an issue in rural or semi-rural areas), you may also pay more for insurance.

- Special coverages you may need, such as a rider to cover business equipment if you work at home; extra coverage riders for jewelry, artwork, guns, and other items valued in excess of policy limits; and additional liability insurance coverage.

- Your own claims history, as well as the rate of claims in your area, will almost always affect your premium. If you have several claims in the space of a few years, your insurance company may not renew your policy. And in some cases, insurance companies have been pulling out of certain areas or even entire states due to high claim rates. In either case, you should contact your lender or another insurance company immediately to arrange other coverage.

Considering all of these factors, the range for homeowner's policy premiums on a house that would cost, for example, $250,000 to rebuild, with fairly straightforward coverage, no riders, and no additional premiums because of past claims or location in an unfavorable fire environment, is quite broad: anywhere from $400 to $2,000 or more per year. So if you have a good claims history, it pays to shop around for all your insurance needs.

Condominium Insurance

Condominium dwellers purchase a hybrid policy since they are responsible for more of the possible loss than a renter is, but less than a homeowner. Otherwise, the provisions and considerations are very similar to homeowner's insurance.

As with any other insurance purchase, always consult a professional insurance advisor to make sure you purchase the right coverages for your needs and meet all state and other requirements.

What About Riders?

As noted above, many items receive only limited coverage under a basic property insurance policy, and many disastrous occurrences are not covered at all. If you have valuable jewelry, collections (coins, guns, stamps), collectibles or antiques, furs, or similar items, you should seriously consider appropriate riders (also called "floaters") to ensure that your valuables are covered for their true "guaranteed replacement" cost. Note that *you actually have to replace an item* to get the replacement cost rather than its depreciated (or actual cash) value, or the policy limit for valuables. You also have to prove the item's age and value, so keep receipts or have appraisals done, and take photos or videos of your valuables. Also, keep the insured value of your property up to date; many insurance policies have an optional inflation rider, and these are well worth the minimal added cost.

This is a good place to remind you that your smaller valuables might benefit from spending the fire season in a safe-deposit box. The receipts, appraisals, and pictures or videos for *all* your valuables should go in there, too.

Finally, if you live in an area that could be flooded in the aftermath of a wildfire due to uncontrolled runoff, you should seriously consider purchasing a flood insurance policy or rider. You do not have to be in a floodplain to be a victim of a flood. Keep in mind that, in most cases, there is a 30-day waiting period after you purchase the policy for coverage to begin. (See the Flood Insurance section at the end of this chapter.)

AFTER A DISASTER

What is Covered? What is Not Covered?
What Losses are Not Insured?

Most insurance policies have a long list of covered and excluded items, in addition to items that are limited in coverage (such as cash, jewelry, guns, etc.). In addition, there are many fine points to understand about water damage. That's why flood insurance is so important. The following table summarizes the items that are covered or excluded in a typical home-owner's insurance policy, but is not all-encompassing.

Typical Included and Excluded Insurance Policy Coverages

Type of Coverage	Included	Excluded	Explanation
Fire and lightning	X		
Explosion	X		
Looting	X		
Smoke	X		Loss must be sudden and accidental
Vandalism	X		
Theft	X		Must not be committed by a resident of the premises
Collapse	X		As long as the cause was covered by the policy
Sudden and accidental discharge or overflow of water from plumbing, heating, or related system or appliance	X		Does not include water from external sources (gutters, floods), freezing, septic fields, wells, etc.
Earth movement		X	Can be earthquakes, rolling boulders, cracking driveways, etc.
Water damage		X	Floods, collected rain water, mudflows, landslides triggered by water, sewer back-ups, etc.
Nuclear hazard		X	
Faulty zoning, planning, etc.		X	Even if it results in flood or earth movements
Failure to protect your property at and after time of loss		X	
Wear and tear		X	

Other Things You Should Know

- Your policy should have a "Declaration Page." This is a short form that sets out the highlights of your insurance coverage—type of building; special coverages, endorsements, and riders; insured value of the property (the estimated cost to rebuild, not the market value); personal property coverage amount; loss of use amount (for additional living expense purposes); and any discounts.

- Your mortgage company will require that it be named as loss payee on the policy and will want a copy of the "Declaration Page." Make sure you have the current address for the company; mortgages get sold with great frequency, so this information may not be current. If you are notified your loan has been sold, it is *your* responsibility to notify your insurance company or agent of the change and related details.

- If you qualify for a discount (newer home, metal/tile roof, sprinkler system, monitored fire detection system, non-smoker, etc.), make sure you stay in compliance. This means you must truly *be* a non-smoker, the fire detection system must actually *be* monitored, and so forth.

- Take note of your deductible. It is a great way to save money on the cost of insurance, but never set it so high you can't afford to cover that amount if a loss occurs.

- If the insurance company does not receive your premium payment by the renewal date on your policy, you usually have a very limited grace period—about 10 days in most cases. After that, the policy lapses and the insurance company does *not* have to reinstate you, even if you later pay the full past-due amount.

Your insurance company may have checklists or other claims forms for you to use, in addition to the initial claims form. If so, these forms make a great starting place, but they are rarely sufficient for your individual needs. The lists below (one for personal property and one for business property) provide a starting point for organizing your claim, damage assessment, and restoration efforts. Always use your insurance company's forms or required format for all claims submissions.

Who Does What? — Personal Property Restoration

Who Does What?	You	Insurance Company	Vendor	Government Agency
Notify Insurance Company of Loss	X			
Protect Property/Contents From Additional Damage/Loss	X		X	
Initial Loss/Damage Assessment	X	X		
Detailed Loss Assessment	X	X	X	X[1]
Turn Off Utilities	X		X	
Photograph Entire Property and Contents	X	X	X[2]	X[1]
Inspect Property to Determine Extent of Damage and Safety of Property/Structures	X	X	X	X[1]
Arrange for Initial Clean-up	X	X[3]		
Clean up Debris, Remove Destroyed or Damaged Contents	X		X	
Perform Needed Demolition	X		X	
Develop Rebuilding Plans	X		X	
Arrange Contractor(s) (Vendors) for Rebuilding/Restoration	X	X[3]		
Supervise Contractors (Vendors)	X			
Arrange for Subcontractors (Vendors)	X		X[2]	
Supervise Subcontractors (Vendors)	X		X[2]	
Choose Materials (Cabinets, Floor Coverings, Paint, etc.)	X	X[4]	X[2]	
Install Materials			X	
Clean-up			X	
Pay Additional Living Expenses	X	X		
Replace Contents	X	X[4]		
Pay the Bills—You are Ultimately Responsible	X	X[3]		X[5]

[1] Only if there is no insurance, insufficient insurance, or other unusual circumstances.
[2] Vendors may take certain actions to protect themselves from liability.
[3] Your insurance company may or may not help arrange this.
[4] Your insurance company has input here since it is obligated only to restore your property to the condition it was in before the event. If you want extras or upgrades, you have to negotiate and pay for them.
[5] If you qualify for FEMA, Red Cross, SBA, or other assistance.

Who Does What? — Business Property Restoration

Who Does What?	You	Insurance Company	Vendor	Gov't Agency	Landlord[1]
Notify Insurance Company of Loss	X				X
Protect Property/Contents From Additional Damage/Loss	X		X		X
Initial Loss/Damage Assessment	X	X			X
Detailed Loss Assessment	X	X	X	X[2]	X
Turn Off Utilities	X		X		X
Photograph Entire Property and Contents	X	X	X[3]	X[2]	X
Inspect Property to Determine Extent of Damage and Safety of Property/Structures	X	X	X	X[2]	X
Arrange for Initial Clean-up	X	X[4]			X
Clean up Debris, Remove Destroyed or Damaged Contents			X		
Perform Needed Demolition			X		
Develop Rebuilding Plans	X		X		
Arrange Contractor(s) (Vendors) for Rebuilding/Restoration	X	X[4]			X
Supervise Contractors (Vendors)	X				X
Arrange for Subcontractors (Vendors)	X		X[3]		X
Supervise Subcontractors (Vendors)	X		X[3]		X
Choose Materials (Cabinets, Floor Coverings, Paint, etc.)	X	X[5]	X[3]		X
Install Materials			X		
Clean-up			X		X
Replace Contents	X	X[5]			
Pay the Bills—You and/or Your Landlord are Ultimately Responsible	X	X[4]		X	X

[1] Many responsibilities fall to the landlord if you do not own the business property (or to you if you are the landlord).

2 Only if there is no insurance, insufficient insurance, or other unusual circumstances.
3 Vendors may take certain actions to protect themselves from liability.
4 Your insurance company may or may not help arrange this.
5 Your insurance company has input here since it is obligated only to restore your property to the condition it was in before the event. If you want extras or upgrades, you have to negotiate and pay for them.

The following table provides an overview of the coverage available under various types of policies. But it is not all-inclusive, and your actual coverages will vary based on your state's laws and other considerations. Always discuss your needs with an insurance advisor licensed in your state.

Which Policy Covers What?

	Homeowners Renters or Condo	Automobile	Business	Fire Policy or Rider	Flood
Fire Damage to Personal Property in Dwelling	X				
Fire Damage to Personal Dwelling	X				
Fire Damage to Non-Owner Occupied Dwelling				X	
Flood Damage to Personal Property in Dwelling					X
Flood Damage to Personal Dwelling					X
Flood Damage to Non-Owner Occupied Dwelling					X
Fire Damage to Personal Items in Auto	X				
Fire Damage to Personal Auto		X			
Temporary Living Expenses	X				
Fire Damage to Outbuildings, Trees, and Landscaping	X			X	

	Homeowners Renters or Condo	Automobile	Business	Fire Policy or Rider	Flood
Flood Damage to Outbuildings, Trees, and Landscaping					
Fire Damage to Business Property or Contents			X		
Flood Damage to Business Property or Contents					X

(Note: "Earth Movement" is purposely omitted from this table).

FLOOD INSURANCE

A typical homeowner's insurance policy doesn't cover flood damage from external water sources. However, floods, mudflows, and landslides are likely to occur—adding insult to injury—in areas that have been stricken by wildfires. Consider what happened when the Hayman and Buffalo Creek fires in Colorado were finally brought under control and then put out. The rains soon followed. The baked and barren ground failed to absorb the runoff. Creeks and streams overflowed. Flood waters inundated homes and other structures. The whole town of Buffalo Creek was nearly destroyed. Yet, few people even think about floods or flood insurance if they don't live in a designated floodplain.

Who Needs Flood Insurance?

You need flood insurance, and probably already have it, if your home or business is located in a designated floodplain or "Special Flood Hazard Area" (SFHA). In fact, to get secured financing to buy, build, or improve a structure in an SFHA, federal law requires that you purchase flood insurance. Accordingly, lenders that are federally regulated or insured must determine if the structure is located in an SFHA and must provide written notice to the borrower of the requirement for flood insurance. So if you still have a mortgage on your property, you're probably already covered.

But even if you do not live in a floodplain you should consider getting flood insurance. According to the Federal Insurance and Mitigation Administration (a component of FEMA), "All areas are susceptible to flooding, although to varying degrees. In fact, 25% of all flood claims occur in the low-to-moderate risk areas. Flooding can be caused by heavy rains, melting snow, by inadequate drainage systems, failed protective devices such as levees and dams, as well as by tropical storms and hurricanes." (From http://www.fema.gov/nfip/whonfip.htm.) And as suggested by the floods following the Hayman and Buffalo Creek fires, if your home is located in an area prone to wildfires—especially if it is near a creek, river, or other water source, or even a *normally* dry wash—you need flood insurance too.

How Do I Get Flood Insurance and When Does It Take Effect?

You should contact your homeowner's insurance company as soon as possible after a fire starts in your area and ask about getting flood coverage. If your agent or company does not handle flood insurance, you can call the National Flood Insurance Program (NFIP) at 1-888-FLOOD29 or TDD# 1-800-427-5593 or visit the NFIP website at http://www.fema.gov/nfipInsurance/search_results.jsp to obtain a current list of authorized insurers and agents.

It's also a good idea to verify your needs and review your individual circumstances with your insurance agent, or with a legal or financial advisor before purchasing a particular flood insurance policy. If you are not sure you qualify for a particular program, call your state emergency management agency or go to their website for details. Your tax advisor may also have current information since there may be tax breaks available.

Keep in mind that there is a 30-day waiting period before a newly purchased flood insurance policy takes effect. Ask your insurance agent for a written statement showing the effective date. If you have held the policy for fewer that 30 days and you become a flood victim, the *policy will not pay, unless* you were required to buy it to obtain your mortgage. This 30-day waiting period is a result of the floods that occurred in the upper Midwest some years ago. Thousands of homeowners, threatened by rising rivers, rushed to buy insurance just days before the waters demolished their houses and destroyed their personal property. Just as with fire coverage, insurance companies are understandably reluctant to issue policies when the covered loss is imminent.

How Do I Pay For Flood Insurance and What Does It Cost?

You pay for flood insurance the same as for most other insurance: directly to your insurance company. Or, the payment may be included with your mortgage PITI and paid from your escrow account. Contact your lender or insurance agent for details.

Flood insurance is issued on a guaranteed basis, and because the rates are set by the federal government, it costs the same regardless of where you buy it. Deductibles are much like those for your regular homeowner's insurance. You must buy insurance on the structure, but insurance on the contents is optional. If you live in a condo, your condo association buys the structural insurance and you buy the contents insurance.

Sample Flood Insurance Premiums and Coverage

Building Coverage/Regular Program

Occupancy Type	Coverage	Sample Premium*
Single family	$125,000	$593
Two to four family	$150,000	$650
Other residential	$200,000	$1,840
Non-residential	$250,000	$2,700

* Premium values are based on Pre-FIRM (Flood Insurance Rate Map) Special Flood Hazard Area rates and include federal Policy Fee & Expense Constant. Premiums do not include ICC (Increased Cost of Compliance) premium. Data current as of May 1, 2001. Source: FEMA and the NFIP website at http://www.fema.gov/nfip/avgcost.htm.

What Kinds of Flood Insurance Policies are Available?

In the insurance industry, your homeowner's, automobile, liability, commercial, and other property insurance are issued using several "standard forms" that give a description of what is covered and include other important information (such as what is not covered). Flood insurance works the same way, and the coverages and policy forms are the same regardless of the issuing company. There are three standard flood insurance forms:

- Dwelling Policy Form
- General Property Policy Form
- Residential Condominium Building Association Policy Form

Dwelling Policy Form

If you live in or own a single family home, one- to four-unit dwelling, townhouse or row house, condominium, manufactured, or mobile home, this is the policy form you will probably need. This policy covers your structure (if applicable) and its contents. If you rent out the structure to someone else, this policy covers the structure, but the tenant must obtain separate contents coverage. If you live in a unit, such as a condominium, where the association is responsible for the outside, the association's condominium policy, as mentioned below, covers the structures; your dwelling policy form covers your own belongings.

General Policy Form

This form is used by cooperatives and certain timeshares.

Residential Condominium Building Association Policy Form

This is the policy form used by residential condominium building associations to cover the entire building under one policy and covers all units. It does not protect an individual owner from loss to personal property owned exclusively by the unit owner. Improvements within the units and personal property owned in common are or should be covered with individual contents policies. This form may also apply to certain timeshares.

Always check with a properly licensed insurance agent to ensure you purchase the appropriate coverages. The statements contained in this chapter are for informational purposes only and may not apply to your particular situation.

How Much Coverage Can I Buy?

The chart that follows shows the coverage available for different types of buildings. If you own multiple buildings, you need to consult with your agent about coverage. There are separate coverage limits for buildings and contents (a consideration if you own a condo or occupy a rental unit), and there are two programs: Emergency and Regular.[31] For specific information about these programs and which (if either) you are eligible and suited for, contact a licensed insurance agent in your state.

---

[31] The NFIP Emergency Program is the initial phase of a community's participation in the NFIP and was designed to provide a limited amount of insurance at less than actuarial rates. Under the Regular Program, more comprehensive floodplain management requirements are imposed on the community in exchange for higher amounts of flood insurance coverage. See http://www.fema.gov/nfip/intnfip.htm for more information.

Building Coverage

Type of Building	Emergency Program	Regular Program
Single family dwelling	35,000	250,000
2- to 4-family dwelling	35,000	250,000
Other residential	100,000	250,000
Non-residential	100,000	500,000

Source: FEMA and the NFIP website at http://www.fema.gov/nfip/c_cov.htm.

Contents Coverage

Type of Building	Emergency Program	Regular Program
Residential	10,000	100,000
Non-residential	100,000	500,000

Source: FEMA and the NFIP website at http://www.fema.gov/nfip/c_cov.htm.

The National Flood Insurance Program encourages everyone to purchase both building and contents coverage to ensure the maximum possible protection against loss. It also makes the process of sorting out which policy covers which loss or damage much less complicated and your claims may get paid more quickly.

What Does a Flood Insurance Policy Cover? What Does It Exclude?

According to the National Flood Insurance Program website, the Standard Flood Insurance Policy forms contain complete definitions of the coverages they provide. Direct physical losses caused by floods are covered. Also covered are losses resulting from erosion caused by unusually high waves or water currents, as well as flooding caused by severe storms, flash floods, or abnormal tidal surges. Damage caused by mudslides (i.e., mudflows), as specifically defined in the policy forms, is also covered.[32]

The following items are covered under building coverage, as long as they are connected to a power source and installed in their functioning location:

• Sump pumps.

• Well water tanks and pumps, cisterns and the water in them.

---

[32] From http://www.fema.gov/nfip/c_cov.htm.

- Oil tanks and the oil in them, natural gas tanks and the gas in them.
- Pumps and/or tanks used in conjunction with solar energy.
- Furnaces, hot water heaters, air conditioners, and heat pumps.
- Electrical junction and circuit breaker boxes, and required utility connections.
- Foundation elements.
- Stairways, staircases, elevators and dumbwaiters.
- Unpainted drywall and sheet rock walls and ceilings, including fiberglass insulation.
- Cleanup.

Clothes washers, clothes dryers, and food freezers, as well as the food in them, are also covered under the optional contents coverage.

However, basements are largely excluded from coverage. The NFIP defines a basement as any area of a building with a floor that is below ground level on all sides. While flood insurance does not cover basement improvements, such as finished walls, floors or ceilings, or personal belongings that may be kept in a basement, such as furniture and other contents, it does cover structural elements, essential equipment and other basic items normally located in a basement. Many of these items are covered under building coverage, and some are covered under contents coverage.

Flood insurance pays even when no disaster is declared. Statistically, federal disaster declarations are issued in less than 50 percent of flooding incidents. An NFIP policy will pay for flood damage whether or not there is a federal disaster declaration.

What Should I Do When a Flood Threatens My Home?

Flood insurance will help you recover if flood waters damage or destroy your home. But you should be aware of what to do when a flood is imminent to protect yourself and your family and to minimize your losses. The following article appears on the National Flood Insurance Program website and offers some good advice:

Coping With a Flood: Before, During and After

Nobody can stop a flood. But if you are faced with one, there are actions you can take to protect your family and keep your property losses to a minimum.

The most important thing is to make sure your family is safe.

Before a Flood

- Keep a battery-powered radio tuned to a local station, and follow emergency instructions.

- If the waters start to rise inside your house before you have evacuated, retreat to the second floor, the attic, and if necessary, the roof. Take dry clothing, a flashlight and a portable radio with you. Then, wait for help. Don't try to swim to safety; wait for rescuers to come to you.

If Time Permits, Here are Other Steps That You Can Take Before the Flood Waters Come

- Turn off all utilities at the main power switch and close the main gas valve if evacuation appears necessary.

- Move valuables, such as papers, furs, jewelry, and clothing to upper floors or higher elevations.

- Fill bathtubs, sinks and plastic soda bottles with clean water. Sanitize the sinks and tubs first by using bleach. Rinse, then fill with clean water.

- Bring outdoor possessions, such as lawn furniture, grills and trash cans inside, or tie them down securely.

Once the Flood Arrives

- Do not drive through a flooded area. If you come upon a flooded road, turn around and go another way. More people drown in their cars than anywhere else.

- Do not walk through flooded areas. As little as six inches of moving water can knock you off your feet.

- Stay away from downed power lines and electrical wires. Electrocution is another major source of deaths in floods. Electric current passes easily through water.

- Look out for animals—especially snakes. Animals lose their homes in floods, too. They may seek shelter in yours.

After the Flood

- If your home, apartment or business has suffered damage, call the insurance company or agent who handles your flood insurance policy right away to file a claim.

- Before entering a building, check for structural damage. Don't go in if there is any chance of the building collapsing.

- Upon entering the building, do not use matches, cigarette lighters or any other open flames, since gas may be trapped inside. Instead, use a flashlight to light your way.

- Keep power off until an electrician has inspected your system for safety.

- Flood waters pick up sewage and chemicals from roads, farms and factories. If your home has been flooded, protect your family's health by cleaning up your house right away. Throw out foods and medicines that may have come into contact with flood water.

- Until local authorities proclaim your water supply to be safe, boil water for drinking and food preparation vigorously for five minutes before using.

- Be careful walking around. After a flood, steps and floors are often slippery with mud and covered with debris, including nails and broken glass.

- Take steps to reduce your risk of future floods. Make sure to follow local building codes and ordinances when rebuilding, and use flood-resistant materials and techniques to protect yourself and your property from future flood damage.

One of the most important things that you can do to protect your home and family before a flood is to purchase a flood insurance policy. You can obtain one through your insurance company or agent. Flood insurance is guaranteed through the National Flood Insurance Program (NFIP), administered by the Federal Emergency Management Agency. Your homeowners insurance does not cover flood damage.

Don't wait until a flood is coming to purchase your policy. It normally takes 30 days after purchase for a flood insurance policy to go into effect.[33]

---

[33] From http://www.fema.gov/nfip/coping.htm.

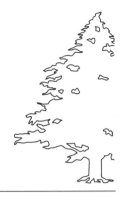

Chapter 6:
History and Background

The following information is from "Federal Wildland Fire Policy" (January 2001), http://www.fs.fed.us/land/wdfire.htm, drafted by the Interagency Federal Wildland Fire Policy Review Working Group, which was composed of representatives from over a dozen federal departments and agencies, including the USDA Forest Service, the Bureau of Land Management, the National Park Service, and FEMA. It provides an overview of the history of wildland fires in the United States and the development of wildland firefighting policy.

OVERVIEW

Role of Wildland Fire in Resource Management

Historical Perspective

"Long before humans arrived in North America, there was fire. It came with the first lightning strike and will remain forever. Unlike earthquakes, tornados, and wind, fire is a disturbance that depends upon complex physical, chemical, and biological relationships. Wildland fire is inherently neither good nor bad, but it is the most powerful natural force that people have learned to use. As an inevitable natural force, it is sometimes unpredictable and potentially destructive and, along with human activities, has shaped ecosystems throughout time.

"Early ecologists recognized the presence of *disturbance* but focused on the principle that the land continued to move toward a stable or equilibrium condition. Through the years, however, scientists have acknowledged that equilibrium conditions are largely the exception and disturbance is generally the rule. Natural forces have affected and defined landscapes throughout time. Inasmuch as humans cannot completely control or eliminate these disturbances, ecosystems will continue to change.

"Human activities also influence ecosystem change. American Indian Tribes actively used fire in prehistoric and historic times to alter vegetation

119

patterns. In short, people and ecosystems evolved with the presence of fire. This human influence shifted after European settlement in North America, when it was believed that fire, unlike other natural disturbance phenomena, could and should be controlled. For many years fire was aggressively excluded to protect both public and private investments and to prevent what was considered the destruction of forests, savannahs, shrublands, and grasslands. While the destructive, potentially deadly side of fire was obvious and immediate, changes and risks resulting from these fire exclusion efforts were difficult to recognize and mounted slowly and inconspicuously over many decades.

Current Perspective

"There is growing recognition that past land-use practices, combined with the effects of fire exclusion, can result in heavy accumulations of dead vegetation, altered fuel arrangement, and changes in vegetative structure and composition. When dead fallen material (including tree boles, tree and shrub branches, leaves, and decaying organic matter) accumulates on the ground, it increases fuel quantity and creates a continuous arrangement of fuel. When this occurs, surface fires may ignite more quickly, burn with greater intensity, and spread more rapidly and extensively than in the past. On the other hand, uses such as grazing can sometimes reduce fine fuels, precluding periodic surface fires that would typically burn in these areas. Without fire, encroachment of woody species may occur in some savannah and grassland ecosystems.

"The arrangement of live vegetation also affects the way fires burn. For example, an increase in the density of small trees creates a multi-storied forest structure with a continuous vertical fuel arrangement. This arrangement may allow a fire normally restricted to the surface to spread into the trees and become a crown fire. In addition to structural changes, vegetation modification resulting from fire exclusion can cause a shift toward species that are not adapted to fire (some of which are not native) and are therefore more susceptible to damage from fire. Fire exclusion sometimes favors non-native species in some fire-dependent areas, while in other areas fires may encourage non-native species. Fires in areas of altered vegetation and fuels can adversely affect other important forces within the ecosystem, such as insects and diseases, wildlife populations, hydrologic processes, soil structure and mineralogy, and nutrient cycling. Any of these components, if altered greatly by unusually severe fire, can seriously diminish the long-term sustainability of the land. In addition, effective protection from, and control of, these large fire events will likely be much more difficult.

"Paradoxically, rather than eliminating fire, exclusion efforts, combined with other land-use practices, have in many places dramatically altered fire regimes (circumstances of fires, including frequency, intensity, and spatial extent) so that today's fires tend to be larger and more severe. No longer a matter of slow accumulation of fuels, today's conditions confront us with the likelihood of more rapid, extensive ecological changes beyond any we have experienced in the past. To address these changes and the challenge they present, we must first understand and accept the role of wildland fire and adopt land management practices that integrate fire as an essential ecosystem process.

"While other techniques, such as mechanical removal, may be used to reduce heavy fuels, they cannot always replace the ecological role that fire plays. Fire not only reduces the build-up of dead and downed fuel, it performs many other critical ecosystem functions. Fire can recycle nutrients that might otherwise be trapped for long periods of time in the dead organic matter that exists in many environments with slow rates of decay. It can also stimulate the production of nutrients and provide the specific conditions, including seed release, soil, light, and nutrients, that are critical for the reproduction of fire-dependent species. ...

* * *

Reintroduction of Fire

"Several factors hinder the reintroduction of wildland fire on an ecologically significant scale. Even now it sometimes takes years to reach agreement about appropriate treatments and to take action. Land managers often feel the need to wait for scientific certainty before acting. This favors the status quo, impedes progress, and deters investigation of new techniques. In some ecosystems, little or no information is available about disturbance regimes, historical fire patterns, response to past management actions, and likely future responses. Information needed to reintroduce fire includes a well-planned, large-scale scientific assessment of current ecosystem conditions and the consequences of various management strategies.

"Another constraint is that Fire Management Plans are not in place in all areas, thus precluding managers from taking advantage of the management options presented by wildland fires. Planning should consider all wildland fires, regardless of ignition source, as opportunities to meet management objectives. In areas where planning has determined a range of appropriate management actions for the use of wildland fire, there will be more opportunities to safely and cost-effectively reintroduce fire. This approach will also make suppression resources available for the highest-priority situations. All wildland fire management actions will continue to be based on values to be protected, fire and land management objectives, and environmental conditions. In many situations, such as fires occurring in highly developed areas or during particularly severe weather, immediate initial attack and prompt suppression will still be required.

"An additional contributing factor is the increasing human settlement that encroaches upon wildlands (wildland/urban interface). Such development divides and fragments wildlands, making it difficult to apply ecosystem-based management strategies. This increases the risk of escaped fires and generates complaints about smoke and altered scenic values. In these areas, the use of fire may be limited in spatial extent and, even where fire introduction is desirable, progress may be slow.

"Smoke is perceived as a factor that may affect land managers' ability to use larger and more frequent wildland fire for restoration and maintenance of fire-dependent ecosystems. Several Federal air quality programs under the Clean Air Act (CAA) regulate wildland fire emissions. The Environmental Protection Agency (EPA) is required to set air quality standards for pollutants that affect public health. States are then required to submit plans to ensure measures will be taken to meet those air quality standards. Local areas may also develop plans that may be more (but not less) restrictive than State and national standards.

"In areas where air quality standards are violated, measures must be taken to reduce emissions. Emission control measures for fires that are used to meet management objectives include smoke management techniques that minimize and disperse smoke away from smoke-sensitive areas. Smoke from fires may also cause standards to be exceeded in communities miles away from the source. Currently, prescribed fires are not considered to be a significant cause of nonattainment, but with increased burning to reduce fuels and restore or maintain ecosystem health, this may change. In many areas, fire managers and local air quality authorities have successfully worked together to accomplish fire and land management objectives, resolve conflicts with smoke emissions, and avoid violation of air quality standards. With guidance from the national level to provide consistent interpretation, further cooperation at the local level will help to achieve a balance of air quality and other ecosystem goals.

"Fire is a unique tool that land managers can use to complement agency missions and land management objectives. But in order to successfully integrate fire into natural resource management, informed managers, partners, and the public must build upon sound scientific principles and social values. Research programs must be developed to create this foundation of sound scientific principles. Before fire is applied on an ecosystem-scale, an understanding of historical fire regimes, as well as a knowledge of the current conditions of each system, is needed. Then all parties must work together in the land management planning and implementation process according to agreed-upon goals for public welfare and the health of the land.

Education

"For many people, fire remains a fearsome, destructive force that can and should be controlled at all costs. Smokey Bear's simple, time-honored "only you" fire prevention message has been so successful that any complex talk about the healthy, natural role of fire and the scientific concepts that support it are often lost by internal and external audiences. A comprehensive message is needed that clearly conveys the desired balance of avoiding fires with adverse effects while simultaneously increasing ecologically beneficial fire.

"The ecological and societal risks of using and excluding fire have not been adequately clarified and quantified to allow open and thorough discussions among managers and the public. Few understand that integrating fire into land management is not a one-time, immediate fix but a continual, long-term process. It is not an end in itself but rather a means to a more healthy end. Full agency commitment to internal and external information and education regarding fire and other ecological processes is needed. Adaptive and innovative fire and land management is severely limited when agency employees and the public misunderstand or remain skeptical about the role of fire.

The Task

"The task before us—reintroducing fire—is both urgent and enormous. Conditions on millions of acres of wildlands increase the probability of large, intense fires beyond any scale we have witnessed. These severe fires will in turn increase the risk to humans, to property, and to the land upon which our social and economic well-being is so intimately intertwined."

ILLUSTRATIONS

The following photographs, which appeared in a White House report titled "Healthy Forests: An Initiative for Wildfire Prevention and Stronger Communities (August 22, 2002),[34] are instructive of the issues.

---

[34] See http://www.whitehouse.gov/infocus/healthyforests/Healthy_Forests_v2.pdf.

Bitterroot National Forest 1895:
Natural Forest Conditions

The 1895 photo shows natural forest stand conditions that evolved from regularly occurring, low-intensity, surface burning. The forest was open and dominated by fire-tolerant, fire-adapted ponderosa pine.

Bitterroot National Forest 1980
Unmanaged Forest

The 1980 photo (from the same place) shows how the forest has changed dramatically since 1895. Over the years small trees have established into dense thickets. These fire-intolerant tree species now crowd the forest, pre-disposing the area to insect infestations, disease outbreaks, and catastrophic wildfires.

Bitterroot National Forest 2001
Catastrophic Wildfire
In this 2001 photo (again, from same place) no "forest" and only a few trees survived the severe fire. Note the beginning of erosion in the stream channel. (The house had been moved prior to the fire. However, this is seldom an option for residents.)

PRESCRIBED BURNS AND USE OF FIRE

The "Federal Wildland Fire Policy" (January 2001), http://www.fs.fed.us/land/wdfire.htm, also states the following with regard to prescribed burns and the use of natural fires:

Use of Wildland Fire

Background

"The use of wildland fire to accomplish land and resource management objectives is referred to as prescribed fire, the deliberate application of fire to wildlands to achieve specific resource management objectives. Prescribed fires may be ignited either by resource managers or by natural events such as lightning. Wildland fire may be used to accomplish a number of resource management purposes, from the reduction of fuel hazards to achieving specific responses from fire-dependent plant species, such as the regeneration of aspen. Often, multiple fire protection and resource management benefits are achieved concurrently.

"Prescribed burning is a well-established practice utilized by public and private land managers. In order to effectively use prescribed fire, land managers must prepare comprehensive burn plans. Each plan specifies desired fire effects; weather conditions that will result in acceptable fire behavior; and the forces needed to ignite, hold, monitor, and extinguish the fire. Generally, the practice of prescribed burning has been used on a relatively small scale and confined to single land ownerships or jurisdictions. Success has been built around qualified and experienced people, their understanding of plant communities and terrain conducive to the use of fire, adequate funding, a supportive public, and a willingness on the part of agency administrators to assume a reasonable amount of risk to achieve desired results.

"Recent fire tragedies in the West have helped to focus attention on the need to reduce hazardous fuel accumulations. Many areas are in need of immediate treatment of both live and dead vegetation to prevent large-scale, high-intensity fires and to maintain their sustainability as healthy ecosystems. Fuel treatment may be achieved by mechanical, chemical, biological, and manual methods, including the use of fire. Strategic landscape-scale fuel management and fire use planning, often integrating a variety of treatment methods, will be necessary to cost-effectively reduce fuel hazards to acceptable levels and to achieve both ecosystem health and resource benefits. Both naturally occurring fuels and hazardous fuel accumulations resulting from resource management and land use activities must be addressed.

Implementation

"Managing for landscape health requires expansion of cooperative interagency prescribed fire programs. Agencies must make a commitment with highly qualified people, from leader to practitioner, and provide funding mechanisms to conduct the program. Federal agencies must foster a work force that understands the role of fire and, at the same time, raise the level of public understanding. Public opinion and perception may limit increases in interagency prescribed fire programs if this is not achieved. Therefore, continued Federal efforts to work collaboratively with and educate private landowners, interest groups, and the media is paramount. Education efforts should focus on exposing the public to accurate information on the environmental, social, and economic benefits that result when prescribed fire is used; how natural resources may be maintained; and the risks involved, including those associated with not taking any action. Increased use of wildland fire may also increase public exposure to smoke and reduced visibility. Understanding of the trade-offs involved is an important educational objective.

"Recent concerns about potential climate change caused by increased carbon dioxide in the atmosphere have also raised questions about the potential impacts of increasing the use of fire. Current analysis suggests that the carbon dioxide released from prescribed fires is ultimately removed by the subsequent regrowth of vegetation. Lower-intensity prescribed fires emit far less carbon dioxide than high-intensity fires. Therefore, if the occurrence of high-intensity fires is reduced through an increase in prescribed burning, a net reduction in carbon dioxide emissions will be achieved. On the other hand, the effects of global warming and increased carbon dioxide on fire occurrence are still being determined. Possibilities include higher rates of fuel accumulation and a warmer climate with more days that favor the occurrence of wildland fire. This may mean it is even more important to increase the use of fire for ecosystem management and hazard fuel reduction. The policies described in this report are consistent with current concerns about climate change. In any case, information about changes in the atmosphere should be incorporated into the preplanning required by these policies.

Administrative Barriers

"In the current atmosphere of downsizing and reduced budgets, agencies may not be able to maintain sufficient numbers of qualified personnel to accomplish broad-scale prescribed fire programs. Many of the employees who are most experienced in the application of prescribed fire are the same employees who are responsible for wildfire suppression. This can lead to competition for their time during the fire season. Administrative procedures also inhibit temporary hiring of personnel needed to conduct on-the-ground prescribed burning activities.

* * *

"To provide optimal biological benefit to forests and rangelands, the timing and intensity of prescribed fire used for ecosystem maintenance should resemble a natural occurrence. Historically, fires were often very large; however, current land-ownership patterns, development, and the processes of funding and conducting prescribed fire are not conducive to replicating this process. For example, it is difficult to have a landscape-size project without involving lands of another ownership, and there are barriers to spending agency funds on non-agency lands. Further, planning, budgeting, and accomplishment reporting processes do not encourage managers to plan large projects with multiple benefits, even when located entirely on agency-administered lands.

"Lastly, there is no consistent method to determine the potential for a prescribed fire to escape, nor is there a mechanism to compare the values at risk from an escaped fire versus those at risk by continuing to exclude fire. When a prescribed fire does escape, the only way a private property owner can be compensated for more than $2,500 in damages is to pursue a tort claim against the Federal government. To prevail, the damaged party must prove negligence on the part of the agency. This cumbersome process leads to ill will between the managing agency and neighboring landowners, adversely affecting cooperation.

Risk Management

"Because of the potential for unintended consequences, prescribed fire is one of the highest-risk activities that Federal land management agencies engage in. Escaped prescribed fires can result from poorly designed or poorly executed projects; they can also result from events beyond the control of those conducting the project, such as unpredicted winds or equipment failure. Currently, the stigma associated with an escaped prescribed fire does not distinguish between poor performance and an unfortunate consequence of unplanned events.

"Although fire is used to accomplish resource objectives in many areas of the United States, other than in the South it is rarely used enough to improve ecosystem health or to reduce fuel hazards on a landscape scale. One reason for this is a lack of commitment to the use of fire. While land management agencies as a whole generally recognize the role of fire as a natural process, not all individual disciplines and managers fully understand or support this role. Some managers are unwilling to accept the risk of potential negative consequences associated with prescribed fire. Differences of opinion concerning the effect of fire on specific resources, such as cultural resources, water quality, air quality, and certain flora and fauna, can also impede the use of fire as a management tool."

FEDERAL WILDLAND FIRE POLICY GUIDING PRINCIPLES

The 2001 Federal Wildland Fire Management Policy[35] states that current federal policy is guided by the following principles:

1. **Firefighter and public safety is the first priority in every fire management activity.**

2. **The role of wildland fire as an essential ecological process and natural change agent will be incorporated into the planning process.**

---

[35] See http://www.nifc.gov/fire_policy/docs/chp3.pdf.

Federal agency land and resource management plans set the objectives for the use and desired future condition of the various public lands.

3. **Fire Management Plans, programs, and activities support land and resource management plans and their implementation.**

4. **Sound risk management is a foundation for all fire management activities.** Risks and uncertainties relating to fire management activities must be understood, analyzed, communicated, and managed as they relate to the cost of either doing or not doing an activity. Net gains to the public benefit will be an important component of decisions.

5. **Fire management programs and activities are economically viable, based upon values to be protected, costs, and land and resource management objectives.** Federal agency administrators are adjusting and reorganizing programs to reduce costs and increase efficiencies. As part of this process, investments in fire management activities must be evaluated against other agency programs in order to effectively accomplish the overall mission, set short- and long-term priorities, and clarify management accountability.

6. **Fire Management Plans and activities are based upon the best available science.** Knowledge and experience are developed among all wildland fire management agencies. An active fire research program combined with interagency collaboration provides the means to make these tools available to all fire managers.

7. **Fire Management Plans and activities incorporate public health and environmental quality considerations.**

8. **Federal, State, tribal, local, interagency, and international coordination and cooperation are essential.** Increasing costs and smaller work forces require that public agencies pool their human resources to successfully deal with the ever-increasing and more complex fire management tasks. Full collaboration among federal agencies and between the federal agencies and international, State, tribal, and local governments and private entities results in a mobile fire management work force available for the full range of public needs.

9. **Standardization of policies and procedures among federal agencies is an ongoing objective.** Consistency of plans and operations provides the fundamental platform upon which federal agencies can cooperate, integrate fire activities across agency boundaries, and provide leadership for cooperation with State, tribal, and local fire management organizations.

SMOKEY BEAR: "ONLY YOU CAN PREVENT WILDFIRES"[36]

We grew up with him, we saw him on countless television ads, he greeted us at the entrances of national parks, national forests, and many other summer vacation destinations, and yet many people don't really understand what Smokey Bear stands for. Too often, he is viewed as a symbol of complete fire suppression and a figure that condemns fires of any type. But the reality is rather different. Smokey Bear was developed and marketed to encourage us to be aware of careless use of fire and the unintended consequences—forest fires, loss of life, and property destruction. He was never intended as a symbol that condemned all fires in our wildlands.

As we learn more about historical fire cycles and their role in the propagation of certain tree species, improving wildlife habitat, and so forth, we are realizing that fire is both a friend and a foe. Some fires can, and must, be allowed to burn to prevent future conflagrations. So, Smokey Bear's new message, "Only YOU Can Prevent Wildfires," carries with it a modern relevance. Some wildfires are started by natural events, but many are caused by careless human activity. What we can do is revisit the history of Smokey Bear and keep in mind the information in the discussion on "Prescribed Burns and Use of Fire" above.

The following quote is interesting and appropriate:

> "For many people, fire remains a fearsome, destructive force that can and should be controlled at all costs. Smokey Bear's simple, time-honored "only you" fire prevention message has been so successful that any complex talk about the healthy, natural role of fire and the scientific concepts that support it are often lost by internal and external audiences. A comprehensive message is needed that clearly conveys the desired balance of avoiding fires with adverse effects while simultaneously increasing ecologically beneficial fire."[37]

Unless you've been living in a cave, you have probably heard of Smokey Bear. Who hasn't heard or read the slogan, "Remember, Only YOU Can Prevent Forest Fires"? But, did you know that in 2001 the

---

[36] Information in this section comes from http://www.smokeybear.com/vault/default.asp.
[37] From the federal Wildland Fire Management Policy & Program Review, Draft Report, June 9, 1995, at http://www.fs.fed.us/land/wdfirex.htm.

slogan was updated to reflect the great risk of fires in wildlands, to "Only YOU Can Prevent Wildfires"? The origins of Smokey Bear and the original message he conveyed are actually quite interesting, and it is the message that has sometimes been misunderstood, as the preceding quote points out. So, here's the true story of the origins of the critter we know as Smokey Bear (yes, it's Smokey Bear, not Smokey THE Bear!).

In 1942, the War Advertising Council needed to encourage North Americans to make a concerted effort to prevent forest fires in order to conserve valuable forest products and reduce the number of people needed to fight the fires (to free them for the war effort). The War Advertising Council developed many slogans, but they were not having the intended effect, so someone came up with plans to use animals to convey their message. Under a one-year agreement with Walt Disney, the first campaign featured a "Bambi" poster (1944). This approach was highly successful, so the Forest Service and the Office of War Information decided to continue a wildfire awareness approach that featured forest animals. They tried several animals, ultimately choosing a bear in 1945. This bear was dressed in a Forest Service uniform and appeared much as he does today. It was decided to name him "Smokey."

It didn't take long for Smokey Bear ("Simon" in Mexico) to become a popular and well-recognized symbol of the continent-wide efforts to prevent human-caused forest fires, and now wildfires. The actual slogan, "Remember, Only YOU Can Prevent Forest Fires" was developed in 1947 by the advertising firm Foote, Cone & Belding for what is now the Advertising Council. Other than updating Smokey Bear from a cub to a mature bear (1948), the message and the medium have remained essentially unchanged.

In 1950, a small brown-phased black bear cub was found after the disastrous Las Tablas fire in New Mexico. He was badly burned and separated from his mother; a local rancher nursed him back to health, and the bear was quickly nicknamed Smokey for the situation in which he was found. When it was suggested to staff of the U.S. Forest Service that this little critter could become a living symbol of the national wildfire prevention program, they quickly agreed. He was transferred to the National Zoo in Washington, D.C. Since no bear lives forever, a successor was found in the Lincoln National Forest in New Mexico (also a brown-phased black bear) in 1971, and brought to the National Zoo. In 1975, the original Smokey Bear retired, and the second bear took over his duties. This living symbol of our continent's wildfire suppression efforts was visited by millions of people each year until his death in August 1989.

Smokey Bear is one of the most widely recognized public service images in the world. He has been adopted across North American (Smokey Bear and Simon). Wildfires have no respect for international borders, so Smokey Bear in the United States and Canada and Simon in Mexico are the cornerstones of an international wildfire prevention program. In the United States, Smokey's image and message are jointly administered by the USDA Forest Service, the National Association of State Foresters, and the Advertising Council. In Canada, Smokey Bear was introduced in 1956 and is managed by the Canadian Forestry Association. Simon in Mexico followed soon after.

One of the most valuable aspects of the Smokey Bear program is its international recognition. He reminds visitors from all over about their responsibility to avoid careless actions that might start or spread a wildfire in the continent's wildlands.

Over time, some people have associated fire prevention with a requirement for 100 percent fire suppression. In reality, the "Only YOU Can Prevent Wildfires" campaign is intended to make people aware of the results of careless actions with fire-starting materials; it was never intended to exclude fire management by trained forestry professionals. As the Canadian Forestry Association website points out:

"It is emphasized that the Smokey Bear campaign against wildfires is intended to reduce the number of fires caused by human carelessness which threaten ... important forests. It is not a one-sided campaign which ignores the larger place of fire management in forestry. Fire has always been a good servant but a poor master. Under special conditions, professional forest managers may actually use fire to reduce fuel hazards in special areas, to improve some wildlife habitats or to assist in the preparation of seedbeds for new forests. But fire—the good servant—is a tool requiring great skill and is employed by foresters only under certain conditions and with a great deal of thought and preparation both before and during its application. Fire—the bad master—is the special target of Smokey Bear because it starts up in places where it may do great harm and can get out of control and threaten valuable forests as well as human lives and settlement ... Smokey Bear's message is simple and reminds us that human carelessness is responsible for most forest wildfires in North America and human care is needed to prevent them. It is up to each of us to make sure that needless damage is not done to our forest land through improper use of campfires, debris burning, matches or smoking material."[38]

Smokey Bear's slogan and ongoing campaign are intended to increase public awareness of the causes of wildfires and the need for proper wildland management, prescribed burns, and controlled use of naturally occurring fires. According to information on a number of websites, the main causes of wildfires caused by human actions or carelessness are:

- Unattended campfires;
- Coals from barbecues;
- Trash burning on windy days;
- Careless disposal of smoking materials or lighted matches; and
- Operating outdoor equipment (all-terrain vehicles, off-road vehicles, chainsaws, etc.) that is not equipped with spark arresters.

To reinforce public awareness of these fire starters, a new campaign has begun to target adults who are casual users of the great outdoors—campers, bikers, hikers—and people who live in the wildland/urban interface. According to the Advertising Council:

> "Wildfire awareness is not a real priority, and most adults feel that they, personally, could never be the cause of a wildfire, despite admitting to occasional carelessness. On the other hand, public awareness of Smokey Bear is high. Market research shows that in terms of recognition, Smokey ranks alongside Santa Claus and Mickey Mouse."[39]

According to FEMA, "More than four out of every five forest fires are started by people. Negligent human behavior such as smoking in forested areas or improperly extinguishing campfires are the cause of many fires."[40] And finally:

> "The USDA Forest Service is responsible for overseeing the use of Smokey Bear in cooperation with the National Association of State Foresters and the Advertising Council. Human caused fire remains the major cause of wildfires across the country. Based on the last 10 years, more than 102,000 wildfires start

---

[38] See http://collections.ic.gc.ca/canforestry/highend/smokeybear/smokeybearhistory.html.
[38] See http://forestry.about.com/library/weekly/aa043001a.htm.
[40] See http://www.fema.gov/hazards/fires/wildlan.shtm.

each year through human carelessness while only 13,000 fires are started by lightning. The Smokey Bear Campaign is a critical tool specially designed to ask for every citizen's conscientious commitment to reduce the expensive resource losses and high suppression costs associated with wildfires. In 1999, more than $500 million was spent suppressing wildfires."[41]

---

[41] See http://forestry.about.com/library/weekly/aa043001a.htm.

AFTERWORD

By Michael G. Apicello

Living With Wildfires provides a snapshot in time of the many faceted social dynamics of wildfire and what it can do today to those people who live in or move close to, high-risk wildfire zones. Living With Wildfires illustrates how a single force of nature can be both beneficial for forests and rangelands and at times, extremely harmful or deadly to humans and communities. The reason for this publication is that no one knows when a wildfire will strike, and quite frankly we want people to know how to be better prepared when they do occur. This book is based on common sense, experience, and science—it is written for the purpose of reducing risks to homes by making them and the landscape around them more survivable when a wildfire does pass through.

Wildfire is a force that also knows no boundaries. It cannot distinguish between what is good to burn and what is not. When manageable "good" wildland fires occur, they can replenish forest soils and promote their health and well-being. Healthy forests provide healthy environments and give us quality products such as clean water, firewood, fiber, habitat, wildlife, and safe places for recreation. But when uncontrolled wildfires get to "running" with thunderous heads of steam, they cannot distinguish what not to burn and can consume everything in their path, including homes and communities. More so, wildfires can easily take down structures where nothing has been done to reduce risk. And where nothing has been done, those structures easily can become another component of available fuels. That's why it is important to make your home *firewise* and to get your community involved in *firewise* activities.

When it comes to being prepared, firefighters appreciate knowing that a home has been treated to reduce surrounding fuels. And when fires strike entire communities, it is those homes that are within the "Firewise Zones" that stand a better chance of survival than those where nothing has been done. In some cases theses structures themselves act as just more fuel in the pathway of a wildfire's rage. Living With Wildfires is as much a basic primer for homeowners as it is a guide to get you on the right track to becoming a member of a "Firewise Community." It is full of information that when followed can lessen the risk of losing your home to wildfire.

Moreover, the information in Living With Wildfires is a noble attempt at capturing the latest fire survival science provided by research and the extensive professional materials written by the experts who manage wildland fires. These are the experts that know how fire behaves and what it can do in different places and at different times of the year. And they offer these experiences in the pages of *Living With Wildfires* with the hope that over time, the information provided will make your home and community safer.

It is no secret across the planet that there are problems with overabundances of fuels in the wildlands. And regardless, homes and communities are being carved into these environments more and more every day. Please take the time not only to read *Living With Wildfires* but to do what it recommends! Make sure your property is as *firewise* as you can. You may be surprised if you follow the Firewise.org guidelines that your structure will still be standing, where others have burned. It's not a guarantee, but the odds will be much greater and the rewards highly worth the effort.

Of course there are no absolute guarantees that even after following all the steps to making your home defendable that it will not burn—but knowing what to do in preparing for escape or evacuation is also very important to know. And sharing this information with friends and neighbors that will help reduce the stress of the unknown when you return home from an evacuation.

For more information, visit **www.firewise.org** or **www.nifc.gov**. These websites reach into the deep network of fire knowledge and lessons learned over time. They are sites for both for the professional firefighter and for homeowners like you. For there is one thing that we all have in common when fire visits our towns—*we are all in this together!*

As surely as the sun will rise and the nights will turn the sky dark, there will always be a time and a season somewhere in the world when wildland fire will race across the landscape. Fire has been a natural element since the beginning of time and will continue to be with us in both good and bad times. But it is important to know when you make the decision to live in the "wildlands" that there are things you can do to make your home as save and as *firewise* as possible.

— Michael G. Apicello[42]
April 2003

---

[42] Michael G. Apicello is the Public Affairs Officer for the National Interagency Fire Center.

GLOSSARY[43]

ACV: Actual Cash Value. The actual cash value is the amount your property or building is worth after it has been depreciated.

ALE: Additional Living Expenses.

Anchor Point: An advantageous location from which to start the construction of a fire line. It is used to minimize the chance of being flanked by fire while the fire line is being constructed.

ARC: American Red Cross.

Ash Pit: A hole filled with hot ash left by burned trees and stumps. Ash pits are dangerous to returning homeowners and firefighters because they may go unnoticed until someone steps into one, causing burns or other injuries. Also called a "stump hole."

Backfire: A firefighting technique in which a fire set along the inner edge of a control line to consume the fuel in the path of a wildfire and/or change the direction of force of the fire's convection column.

Backfiring: A firefighting control technique involving intentionally igniting fuels inside a control line to deprive a wildfire of fuel to burn.

Backing Fire: Fire that is moving into the wind. Backing fires burn with a slower rate of spread and at a lower intensity. Compare Head Fire and Flanking Fire.

BAER Teams: Burned Area Emergency Rehabilitation Teams. This is a team of natural resource and forestry engineering experts who develop and implement plans to help restore burned areas on public and nearby private lands.

Bambi Bucket: A collapsible bucket slung below a helicopter. Used to dip water from a variety of sources for fire suppression.

Barrier: Any obstruction to the spread of fire; can be natural or man-made. Barriers are often roads, or rivers, or lakes.

Black: Refers to areas where a fire has recently burned and consequently no fuel exists. The "black" may be used as an area of safety from wildfire in the event of an emergency, but only if the area is solid or completely black.

Blackline Concept: Refers to a strip adjacent to a control line that is intentionally burned in order to contain a fire. Many prescribed fire

---

[43] Compiled from glossaries posted on several fire agency websites, including the "Glossary of Wildland Fire Terms" by the National Interagency Fire Center.

and suppression techniques are based on the concept of "blackline" as a barrier to wildfire spread. This is a firefighting technique.

BLM: Bureau of Land Management (part of the U.S. Department of the Interior). This is one of many fire authorities and federal government land management entities.

Blowdown: Trees or snags that are blown over by a windstorm. They may often be weakened, over-pruned, or damaged by fire. A live or dead tree whose trunk, root system or branches have deteriorated or been damaged to such an extent as to be a potential danger to human safety.

Blow-up: A sudden increase in fire intensity, or rate of spread strong enough to prevent direct control or to upset control plans. Blow-ups are often accompanied by violent convection and may have other characteristics of a firestorm. (See **Flare-up.**)

Bole: The (inner) bark around the base of a tree, that serves to protect the cambian layer. Often referred to as the trunk of a tree.

BTU: British Thermal Unit (a measurement of heat output).

Bucket Drop: Water dropped by a helicopter enabled with a sling and remotely operated bucket that is filled from a water source. This is a firefighting technique.

Burn Boss: The incident commander for a prescribed burn.

Burn Out: Setting fire inside a control line to widen it or consume fuel between the edge of the fire and the control line. This is a firefighting technique often done by firefighters as they construct a fire line.

Cambian Layer: A very thin layer of cells from which new bark and wood cells are created. The cambian layer is between the woody part of the tree and the bark. If the bark and the cambian layer are completely destroyed, the tree usually dies.

Carrier Fuel: Fuels that allow a fire to spread and "carry" fire through the forest. These are generally lighter fuels such as conifer needles, leaves, cured grass, and small twigs.

Chain: A traditional forestry term equal to 66 feet or 20 meters.

Chaparral: A type of vegetation consisting of stands of dense, spiny shrubs with tough evergreen leaves. Chaparral is found in the Mediterranean region, as well as along the coastlines of California, Oregon, Washington, Chile, southwestern Africa, and southwestern Australia. Fires are common in chaparral because the shrubs store reserves of food in fire resistant roots.

Coal Seam Fire: A fire that burns underground, often for many years, generally in abandoned coal mines.

Complex: Two or more fire incidents located in the same general area, assigned to a single incident commander or unified command.

Comprehensive Auto Insurance: That part of "full-coverage" auto insurance that pays for "other than collision" damages.

Condominium Insurance: A special form of homeowner's insurance available only to condominium owners.

Conduction: Heat transfer method in which heat is transferred from one molecule to another. The physical spread of heat, like the slow burning of a stump, or a slow burning ground fire.

Confine: Firefighting step in which a fire is restricted within determined boundaries.

Containment: Firefighting phase that uses fire suppression action to check the spread of a fire under prevailing conditions. Usually, a fuel break has been completed around the fire. This break may include natural barriers and/or a manually and/or mechanically constructed line.

Control: Firefighting phase in which a fire is extinguished enough along the perimeter so that there is no danger of the fire escaping. However, it is still possible to see flames or smoke in the center of the controlled area for several weeks.

Control Line: A line in from which fuels have been removed. A control line may be dug by a fire crew (a hand line) or by machinery such as a bulldozer (a dozer line). A control line may be established by wetting fuels using engines and fire hoses (a wet line). This is a common approach used by firefighters to protect their structures. A control line may also consist of a road, river, snow bank, rock outcropping, or other barrier to fire spread. This is a firefighting technique.

Convection: Heat transfer method in which heat is transferred by air movement. The hot, rising gases often cause fire to burn upslope.

Cover Type: A standardized description of the vegetation in an area in which a fire is burning. The cover type is based on the dominant tree species and the age of the forest. Cover types help predict site moisture, fuel loading, and fire behavior. (See **Vegetation**.)

Crown Fire (Crowning): A fire that advances top-to-top through trees or shrubs, more or less independently of a surface or ground fire.

Creeping Fire: A low-intensity fire with a low rate of spread.

Deadfall: The collection of dead vegetation, dropped branches, leaves, stumps, needles, and other material that makes up "fuel" in a forest.

Deductible: The amount of risk an insured assumes—the amount of a covered loss the insured will pay before the insurance company begins paying claimed amounts.

Defensible Space: The area around a structure, typically 30 feet wide or more, where dry brush, vegetation, other combustible materials, and obstacles have been removed, giving firefighters room to fight a fire safely. Also, an area either natural or manmade where material capable of causing a fire to spread has been treated, cleared, reduced, or changed to act as a barrier between an advancing wildland fire and the loss to life, property, or resources. May also be referred to as "hazard mitigation zone," "home ignition zone," and "survivable space."

Defensive Mode: Wildland fire fighting mode in which firefighters take a stance to protect structures. In this mode, the movement of the fire is not controlled.

Demonstration Forest: A project area that has been specially thinned to approximate the appearance of a natural Pre-European Settlement time area—with minimal fuels—to show what a forest subject to low intensity fire would look like. This contrasts with the surrounding forest, showing the effects of constant fire suppression and resultant build-up of fuels and spindly, fire-prone trees.

Direct Attack: Any treatment of fuel, such as by wetting, smothering or chemically quenching the fire or by physically separating burning from unburned fuel. This is a firefighting technique. Also called "getting the fire." Fighting the fire adjacent to the fire's edge.

Diurnal Winds: Winds that occur in a repeating 24-hour pattern. For example, in a valley, as the day heats the wind moves upslope, and as the evening cools the wind moves downslope.

Doghair: Stands of very dense trees—thousands of trees per acre instead of hundreds. These trees are ideal kindling for an intensely burning wildland fire.

Downslope Effect: The warming of airflow as it descends a hill or mountain slope.

Dozer: Any tracked vehicle with a front-mounted blade used for exposing mineral soil. Usually used to build fire lines and to provide access for fire fighting equipment.

Dozer Line: A fire line constructed by the front blade of a dozer.

Drip Torch: An ignition tool that drips a flaming mixture of diesel and gasoline onto the ground. This is a specialized tool used as part of a prescribed burn and/or as a firefighting technique.

Dry Lightning: Lightning from a thunderstorm that gives little or no precipitation. Dry lightning often starts wildfires.

Embers: (1) Smoldering ash of a dying fire; (2) Isolated fires are often started in front of the main fire by the sparks and embers lofted in a smoke column. These fires often cause homes to burn. Also called a "spot fire."

Endorsement: Insurance company lingo for special coverages that are added to your insurance policy. (See **Rider.**)

Extreme Fire Behavior: "Extreme" implies a level of fire behavior characteristics that ordinarily precludes methods of direct control action. One or more of the following is usually involved: high rate of spread, prolific crowning and/or spotting, presence of fire whirls, strong convection column. Predictability is difficult because such fires often exercise some degree of influence on their environment and behave erratically, sometimes dangerously.

FEMA: The Federal Emergency Management Agency.

Firebrands: Burning material thrown in the air. Firebrands may be thrown ahead of a main fire or shower down from a burning snag or tree. These are often associated with crown fires and cause rapid spreading of a fire.

Fire line: A linear fire barrier that is scraped or dug to mineral soil. This is a firefighting technique.

Fireproof: Noncombustible. Able to withstand the effects of fire with little or no burning. However, if the fire is hot enough, even fireproof items can melt or vaporize.

Fire Behavior: The manner in which a fire reacts to the influences of fuel, weather and topography.

Fire Behavior Forecast: Prediction of probable fire behavior, usually prepared by a fire behavior analyst in support of fire suppression or prescribed burning operations.

Fire Extent: The area burned per time period or event.

Fire Front: The part of a fire where continuous flaming combustion is taking place. Unless otherwise specified, the fire front is assumed to be the leading edge of the fire perimeter.

Fire Insurance: An aspect or feature of property insurance.

Fire Intensity: The rate of heat release for an entire fire at a specific point in time, typically measured in BTUs per foot per second.

Fire Management Authorities (or Fire Authorities): The various representatives of federal, state, and local agencies and other entities that are responsible for managing and fighting wildfires, planning prescribed burns, and carrying out other activities related to managing wildfires.

Fire Names: The normal procedure or protocol is that the Incident Commander names a fire. The name is usually taken from a nearby local geographic feature.

Fire-resistant: Something that can withstand a certain amount of fire without too much burning. If the fire is hot enough or lasts long enough, fire-resistant materials will burn or melt.

Fire Retardant: A substance that by chemical or physical action reduces the flammability of fuels or slows the spread of the fire.

Fire Return Interval: The average amount of time between successive fire seasons. Fire return interval may be specified for a single point on the landscape (point fire return interval) or for an area of a specified size.

Fire Use: A fire that was started by natural conditions in a remote area in which a fire plan already exists. These fires are allowed to burn as long as they do not threaten lives, property, or natural resources, and are only fought if they escape predetermined boundaries or environmental conditions change which make it too hazardous to allow the fire to continue.

Firewise Construction: The use of materials and systems in design and construction to safeguard against the spread of fire within a building or from a building to the wildland/urban interface area.

Firewise Landscaping: Landscape design that replaces flammable fuels around a structure with fire-resistant materials and vegetation such as green lawns, gardens, some types of individually spaced green ornamental shrubs, individually spaced and pruned trees, decorative concrete, gravel, brick, stone, and other non-flammable or flame-resistant materials.

Flame Length: The average length of the flame from the ground to the flame tips.

Flanking Fire: Fire that is moving perpendicular to or at an angle to the wind direction. Often considered to be the side of a fire. Its intensity is lower than a Head Fire and higher than a Backing Fire.

Flare-up: Any sudden acceleration of fire spread or intensification of a fire. Unlike a blow-up, a flare-up lasts a relatively short time and does not radically change control plans. A flare-up can be a single tree.

Flood (for insurance purposes under the National Flood Insurance Program): A general and temporary condition of partial or complete inundation of two or more acres of normally dry land area or of two or more properties (at least one of which is the policyholder's property) from: (1) overflow of inland or tidal waters; (2) unusual and rapid accumulation or runoff of surface waters from any source; or (3) mudflow. Also: A collapse or subsidence of land along the shore of a lake or similar body of water as a result of erosion or undermining caused by waves or currents of water exceeding anticipated cyclical levels that result in a flood as defined above.

Floodplain: Generally the area adjacent to or near a river, stream, lake, or other body of water that has been determined by appropriate authorities (generally the Army Corps of Engineers) to be at risk of flooding.

FMA: Fire Management Assistance.

Foam: A surfactant often used for firefighting. This is one of the main ways to protect structures. Foam insulates against heat, increases the penetration of water into fuels, and decreases evaporation.

Foehn (commonly pronounced "fern"): A type of warm, dry wind that occurs when stable, high pressure is forced across and then down a mountain range. Called by various names such as Santa Ana, Chinook, Mono, East, and North Wind.

Fronts: The transition zone between two different air masses.

Fuel: Combustible material available to a fire. Fuels are classified by type according to the amount of time a fuel takes to gain or lose moisture. Fuels may also be referred to as fine (needles, leaves, lichen); live or green (living foliage and branches); downed (fuel on the ground); or heavy (large logs and snags).

Fuel Break: A strip of land where vegetation has been modified or cleared and acts as a buffer to fire spread. Often selected or constructed to protect an area from fire. This area does not have to be barren of plant life, but plants need to be carefully chosen to provide a safe area for firefighters to operate and to minimize the ability of a fire to move from plant to plant or tree to tree.

Fuel Continuum: The spectrum of potential fire fuels, ranging from the finest mulch, leaves, and pine needles, through twigs, branches, deadfall and so forth, to bushes, trees, cars, homes, construction materials and anything else that can burn.

Fuel Loading: The mass of available fuel per unit area (in kilograms per meter or tons per acre).

Fuel Moisture: The amount of water in a fuel sample. The fuel moisture is the proportion of water to dry material. Fire behavior is largely dependent on how much water is in the fuel. The lower the fuel moisture, the faster the fire will burn and the more intense it will be.

Full Replacement Coverage: A feature available with many property insurance policies to ensure your structure and contents are adequately insured to meet company requirements for replacing or rebuilding at current market replacement cost, not the depreciated value.

Going Bare/Going Naked: *Choosing* to *not* have insurance coverage.

Grace Period: The short period following the due date for an insurance premium payment. The insurance company will *normally* accept payments that *arrive* during this window (post-marked is not good enough).

Ground Fire: Fire burning in the ground or through the undergrowth and not reaching into the canopy. The material the fire burns includes all combustible materials lying beneath the ground surface, including roots, deep duff, peat, and rotten buried logs.

Hand Line: A fire line built with hand tools. This is a firefighting technique.

Hazard Fuels: Fuels posing a threat to people or property in the event of a wildfire. These are the materials that need to be cleared well away from any structures.

Holdover: A fire that remains dormant for a considerable time.

Homeowners' Insurance: A combination of fire, hazard, contents, and liability insurance carried by most homeowners to protect their property.

Head Fire: Fire that is moving with the wind. Head fires burn at a higher intensity and have a faster rate of spread than backing fires. Usually considered to be the leading edge of the fire. Compare Backing Fire and Flanking Fire.

Heel: The bottom, or near the start of a fire.

Hotshot Crew: A highly trained fire crew, usually consisting of 20 members, used mainly to build fire lines with hand tools.

Hot Spot: A particularly active part of a fire.

Hotspotting: Reducing or stopping the spread of fire at points of particularly rapid rate of spread or special threat, generally the first step in prompt control, with emphasis on first priorities. This is a firefighting technique.

IFG: Individual and Family Grants for emergency repairs, personal property replacement, and transportation not covered by insurance.

Incident Commander: Person responsible for the overall management of the incident.

Incident Management Team: The incident commander and staff assigned to manage the incident.

Indirect Attack: A method of suppression in which the control line is usually at a distance from the fire and sometimes, as appropriate, intervening fuel is burned out.

Inflation Protector or Guard: Terms used by insurance companies and advisors to describe a policy feature that ensures your coverage meets the requirements for full replacement coverage rather than actual cash value. (See **ACV** and **Full Replacement Coverage.**)

Inversion: Under high pressure and stable air conditions warm air may "cap" cooler air, forming an inversion that traps smoke in valley bottoms, particularly at night (also called temperature inversion).

IRB: Internal Revenue Service Bulletin.

IRS: Internal Revenue Service.

Jackpot: A pocket of heavy fuels (such as downed logs) in an area where the fuel load is otherwise low that may flare up and cause greater fire intensity if a fire gets to it.

Ladder Fuels: Fuels such as branches, shrubs, or an understory layer of trees, arranged or growing in continuous layers from the ground up, that allow a fire to spread (or "ladder") upwards from the ground to the forest canopy.

Landslide: The devastating sweep of rocks, dirt, debris, vegetation, building materials, and so forth that may slide loose after a devastating fire and following rainstorms. The damage may be covered by flood insurance (if the slide was caused by certain events) or by earth movement insurance.

Lookout: A member of a fire crew whose job is to monitor local weather conditions, and to identify and report potential dangers resulting from a change in fire behavior or weather. A lookout is also watching out for the crew's safety. A lookout may also refer to a fire-spotting tower or to the people in those towers.

Low-intensity Fire: A fire that burns close to the ground, moving quickly, at low temperatures. This is the "ideal" fire since it does little serious damage to property and reduces dangerous fuel build-up.

Mature Height: The full adult height of a tree. As a rule of thumb, you need to know the mature height of any trees you plant and place them at least two times the height of the tree (depending on your area) away from any structures.

Mitigation: An action intended to reduce the risk or severity of a fire hazard.

Mop-up: To make a fire safe or reduce residual smoke after the fire has been controlled by extinguishing or removing burning material, felling snags, or moving logs so they will not roll downhill. The last phase in fire fighting.

Mudflow: The mixture of water and dirt (see also **Slurry**) that causes additional devastation in many areas following a wildfire. This damage is normally covered by flood insurance.

Naturalized: A tree, shrub, plant, or ground cover that has adapted to a location where it is not native and is now prevalent.

NFIP: The National Flood Insurance Program administered by the Federal Insurance & Mitigation Administration division of FEMA.

Noncombustible: A material that, when used in the form and under the conditions anticipated, will not aid combustion or add appreciable heat to an ambient fire.

Offensive Mode: Wildland fire fighting mode in which firefighters take aggressive action.

Open Fire: A fire whose source is not contained. A camp stove or cigarette inside a closed space is not an open fire; a cigarette outside or campfire is an open fire.

PDA: Personal Digital Assistant (such as a Palm Pilot or Pocket PC).

Point of Origin: Point where the fire started. The area is often protected as evidence for an investigation.

Polhemus Prescribed Burn: The name given to the burn/burn technique that resulted in a firebreak that reduced the severity of the Hayman Fire.

Preheating: An increase in flammability of fuels as a result of exposure to heat and convective wind ahead of a fire. Can be caused by radiant heat.

Prescribed Fire or Burn: A fire ignited by management actions under predetermined conditions and confined to a specific area to fulfill specific resource management objectives. Prior to ignition a written, approved prescribed fire plan must exist and NEPA (National Environmental Policy Act) requirements must be met.

Prescribed Fire Plan (Burn Plan): Provides the prescribed fire burn boss the information needed to execute an individual prescribed fire.

Prescription: Measurable criteria that define the perimeters under which a prescribed fire will be started or allowed to burn. Prescription criteria may include safety, economic, public health, environmental, weather, air quality, geographic, administrative, social, or legal considerations.

Presidentially Declared Disaster: A disaster that occurs in an area declared by the President to be eligible for federal assistance under the Disaster Relief and Emergency Assistance Act.

Primary Fuel Break Zone: The main area around a home or other structure that is free of ladder fuels and other flammable materials. This area does not have to be barren of plant life, but plantings must be carefully chosen to provide a safe area for firefighters to operate and to minimize the ability of a fire to move from plant to plant.

Property Insurance: An insurance policy that covers residence, other structures, some landscaping, contents, and temporary lodging in case of a covered loss such as that due to fire.

Radiation: Heat transfer method where heat is transferred through a wave motion. Heat is generated by a burned or super-heated item such as that from a brick wall, a tree that is burned on the inside, a metal source, and so forth.

Rate of Spread: The speed at which a flame travels (in feet per second; miles per hour, mph; feet per hour, ft/hr; chains/hour; or kilometers per hour, kph).

Realtor®: A registered trademark used by authorized, licensed real estate agents. Realtors® can be excellent sources of referral for the reconstruction and recovery processes.

Reburn or Rekindle: Subsequent burning of an area in which fire has previously burned but has left fuel that ignites when burning conditions are more favorable.

Red Flag Warning: A term used by fire-weather forecasters to call attention to weather of particular importance to fire behavior. It is usually made as a result of notification from the National Weather Service of high temperatures, low humidity and gusty winds that could result in extreme wildland fire behavior. It is a sign of imminent fire behavior.

Red Flag Watch: A term used by fire-weather forecasters to notify the public of the possible development of a red flag warning event in the near future.

Red-tagged: An unofficial slang term for houses and other structures determined to have insufficient defensible space, inadequate ingress and egress, or that are designed or constructed in such a manner that it would be unsafe to fight a fire on the property.

Relative Humidity: The ratio of absolute to saturation vapor pressure. In other words, the amount of moisture in the air relative to the amount it could actually hold. Fire behavior is affected by, and can be predicted using, relative humidity.

Renter's Insurance: A contents and liability-only insurance to protect a tenant's personal belongings.

Rider: A special addition or exclusion to an insurance policy.

Running (of a fire): The rapid advance of the head of a fire with a marked change in fire line intensity and rate of spread from that noted before and after the advance.

Safety Zone: An area cleared of flammable materials used for escape in the event the line is outflanked or in case a spot fire causes fuels outside the control line to render the line unsafe. A safety zone is an area where you can safely stand, and the fire should go around you. Using firing operations or mechanical removal of vegetation, crews progress so as to maintain a safety zone close at hand allowing the fuels inside the control line to be consumed before going ahead. Safety zones may also be constructed as integral parts of fuel breaks; they are greatly enlarged areas which can be used with relative safety by firefighters and their equipment in the event of a blowup in the vicinity. This also includes the "black" areas. A house can be a safety zone if it has a good fuel break surrounding the structure.

SBA: Small Business Administration.

Serotinous Cones: Reproductive cone structure for certain types of trees. The cones are sealed and require intense heat before they will open and release seeds. The cones are an example of plants adapting to fire.

Slop-over: A fire edge that crosses a control line or natural barrier intended to contain the fire.

Slurry: (1) A mixture of water, debris, mud, rocks, and vegetation carried as part of a mudflow or landslide; (2) It is a fire retardant—it slows fires—and is made of 85 percent water, 10 percent fertilizer, and 5 percent other ingredients (iron oxide, clay, bentonite). It is usually dropped from the air.

Slurry Bomber: An air tanker (fixed wing) used to drop fire retardant or water over a burning or at risk area.

Snag: A standing dead tree. Snags are a hazard to firefighters and homeowners returning to their property because the roots can burn out causing the tree to fall unexpectedly, particularly in windy conditions. Firefighters sometimes call these "widow makers." Burning snags may also throw sparks and firebrands ahead of a fire front, causing spot fires. However, snags are often beneficial as wildlife habitats.

Spot Fire: A fire ignited outside the perimeter of the main fire may be caused by flying sparks, embers, or branches.

Stairstep: A building technique to minimize the danger and instability of a steep hillside or slope.

Stand Replacement Fire: A fire of such intensity and severity that nearly all the trees in a stand are consumed. Forests succeeding a stand replacement fire are generally composed of trees that quickly reestablish and consequently are of uniform age.

Strike Team or Task Force: Crews, engines, standardized equipment, with common communications mobilized together to fight a fire.

Surface Fire: A fire that burns the surface of the ground. Also, burning grass, debris, litter, and small vegetation on the forest floor or in a grassland area.

Tender: A fire fighting truck used to carry water.

Thermal Belt: An area on a mountain that usually experiences very little variation in temperature, has the highest average temperature, and thus, the lowest average relative humidity. Area of trapped hot air in-between cooler layers. Fire intensity may be increased within the thermal belt.

Torching Fire: The ignition and flare-up of a single tree or small group of trees, usually from bottom to top.

Uncontrolled Fire: Any fire that threatens to destroy life, property, or natural resources, and either is not burning within the confines of firebreaks, or is burning with such intensity that it cannot be readily extinguished with ordinary tools commonly available. (See **Wildfire.**)

Upslope Winds: The cooling of an airflow as it ascends a hill or mountain slope. If there is enough moisture and the air is stable, precipitation may form. If the air is unstable, there might be an increased chance of thunderstorm development.

USDA: The U.S. Department of Agriculture—parent agency of the U.S. Forest Service.

WGA: The Western Governors' Association. This group coordinates fire policy for member states. See their website, http://www.westgov.org/wga/initiatives/fire/default.htm, for more information.

Wildfire: An out-of-control wildland fire that threatens to destroy life, property, or natural resources, requiring suppression action. (See **Uncontrolled Fire.**)

Wildland: Forests, brushlands, grasslands, and other areas primarily uninhabited by humans, in which development is essentially non-existent, except for roads, railroads, power lines, and similar transportation facilities. Structures, if any, are widely scattered.

Wildland Fire: Any non-structure fire that occurs in the wildland.

Wildland Fire for Resource Benefit: Naturally ignited fires that are allowed to burn within prescribed parameters in order to maintain the natural ecosystem and its processes. (See **Fire Use.**)

Wildland/Urban Interface (WUI): The area between residential developments or suburbs and public lands, open lands, forests, or ranch lands.

BIBLIOGRAPHY

RECOMMENDED READING & REFERENCES

For additional information on wildland fires and wildfires, please see the following publications:

Government Publications on Wildfire Prevention

"A Collaborative Approach for Reducing Wildland Fire Risks to Communities and the Environment" from the USDA Forest Service (August, 2001), available for free download in PDF at http://www.fireplan.gov/FIRE.REPORT.1.pdf.

"Preparing a House for Wildland Fire Season" checklist from Firewise, available for free download in PDF at http://www.firewise.org/prepare/prepare.pdf.

"Protecting People and Sustaining Resources in Fire-Adapted Ecosystems: A Cohesive Strategy" by the USDA Forest Service (October 13, 2000), available for free download in PDF at http://www.fs.fed.us/publications/2000/cohesive_strategy10132000.pdf.

"Recreation Area Fire Prevention" by the National Wildfire Coordinating Group (March, 1999), available for free download in PDF at http://www.nwcg.gov/pms/pubs/recreati.pdf.

"Wildfire Approaching Checklist" from Firewise, available for free download in PDF at http://www.firewise.org/app/approach.pdf.

"Wildfire Prevention Strategies" by the National Wildfire Coordinating Group (March, 1998), available for free download in PDF at http://www.nwcg.gov/pms/docs/wfprevnttrat.pdf.

Flooding and Flood Insurance Information

"Avoiding Flood Damage: A Checklist for Homeowners" from the Federal Emergency Management Agency (FEMA), available for free download in PDF at http://www.fema.gov/pdf/hazards/flddam.pdf.

"Floods and Flash Floods Fact Sheet" from the Federal Emergency Management Agency (FEMA), available for free download in PDF at http://www.fema.gov/pdf/hazards/floodfs.pdf.

Books on Wildland Fire and Firefighting

America's Fires: Management on Wildlands and Forests by Stephen J. Pyne, from Forest History Society (June 1997); ISBN: 0890300534.

Fire and Ashes: On the Front Lines of American Wildfire by John N. Maclean, from Henry Holt & Company, Inc. (June 2003); ISBN: 0805072128.

Fire in America: A Cultural History of Wildland and Rural Fire by Stephen J. Pyne, William Cronon, from University of Washington Press (March 1997); ISBN: 029597592X.

Fire in Their Eyes: Wildfires and the People Who Fight Them by Karen Magnuson Beil, from Harcourt (April, 1999); ISBN: 0152010424.

Fire on the Mountain: The True Story of the South Canyon Fire by John N. Maclean, from Washington Square Press (August, 2000); ISBN: 0743410386.

Homeowner's Guide to Wildfires: In the Urban Interface by Frederick Rodak, from Wildfire Technologies (July 1991); ASIN: 0963049305.

In Fire's Way: A Practical Guide to Life in the Wildfire Danger Zone by Tom Wolf, from University of New Mexico Press (March 2003); ISBN: 0826320953.

Jumping Fire: A Smokejumper's Memoir of Fighting Wildfire by Murry Taylor, from Harvest Books (June, 2001); ISBN: 0156013975.

On Storm King Mountain: The Legacy ... the Lesson by Linda Pascucci and Ron Pascucci, from 1st Books (December, 1998); ISBN: 1585005029.

Structure Protection in the I-Zone: Focusing Your Wildland Experience for the Urban Interface by George Bradford, from Fire Engineering Book Department (August 2000); ISBN: 0912212950.

Wildfire: A Reader by Alianor True (Editor), Stephen J. Pyne, from Island Press (June, 2001); ISBN: 1559639075.

Wildland Firefighting Practices by Mark Huth, Joseph D. Lowe, Jeanne Mesick, Kasey Young; from Delmar Learning (August, 2000); ISBN: 0766801470.

WEBLIOGRAPHY

ORGANIZATIONS, GOVERNMENT AGENCIES, AND ONLINE RESOURCES

For additional information on wildland fires and wildfires, please see the following websites:

American Red Cross:
http://www.redcross.org/

Bureau of Indian Affairs:
http://www.doi.gov/bureau-indian-affairs.html

Bureau of Land Management fire page:
http://www.fire.blm.gov/index.htm

Federal Emergency Management Agency (FEMA):
http://www.fema.gov/

Federal Wildland Fire Policy:
http://www.fs.fed.us/land/wdfire.htm

Fire Management Today online magazine:
http://www.fs.fed.us/fire/fmt/index.html

Fire resistant landscaping information—links by state:
http://plants.usda.gov/cgi_bin/link_categories.cgi?category=linknative#Ref114

Firewise:
http://www.firewise.org/

High Country News:
http://www.hcn.org/

Joint Fire Science Program—U.S. Department of the Interior and USDA Forest Service:
http://jfsp.nifc.gov/

Mental Health Sanctuary:
http://www.mhsanctuary.com/mh/toll.htm

National Association of State Foresters:
http://www.stateforesters.org/

National Emergency Response Team:
  http://www.nert-usa.org/

National Fire Plan:
  http://www.fireplan.gov/

National Flood Insurance Program:
  http://www.fema.gov/nfip/

National Forest Foundation:
  http://www.natlforests.org/

National Interagency Fire Center:
  http://www.nifc.gov/

National Park Service Fire & Aviation Management page:
  http://www.nps.gov/fire/

National Wildfire Coordinating Group:
  http://www.nwcg.gov/

Natural Resources Conservation Service:
  http://www.nrcs.usda.gov/

Salvation Army USA:
  http://www.salvationarmyusa.org/

Small Business Administration Disaster Assistance page:
  http://www.sba.gov/disaster/

U.S. Department of the Interior:
  http://www.doi.gov/

USDA Forest Service Fire & Aviation Management page:
  http://www.fs.fed.us/fire/

USDA Forest Service:
  http://www.fs.fed.us/

Wildfire History and Ecology from the website "Canyons, Cultures and
Environmental Change: An Introduction to the Land-Use History of the
Colorado Plateau," John D. Grahame and Thomas D. Sisk, editors
(2002):
  http://www.cpluhna.nau.edu/Biota/wildfire.htm

Wildland Firefighters Monument
  http://www.nifc.gov/monument/

Yellowstone in the Afterglow: Lessons from the Fires, a free e-book by Mary Ann Franke of the Yellowstone Center for Resources (2000): http://www.nps.gov/yell/publications/pdfs/fire/htmls/cover.htm

NATIONAL ASSOCIATION OF STATE FORESTERS

From: http://www.stateforesters.org/SFlist.html

(Current as of March 10, 2003)

State Foresters list—This is normally your first source of wildfire mitigation information for your particular state and situation, unless you live inside BLM, FS, or NPS lands.

ALABAMA
http://www.forestry.state.al.us/
AL Forestry Commission
513 Madison Avenue
Montgomery, AL 36130

ALASKA
http://www.dnr.state.ak.us/forestry
AK Division of Forestry
State Forester's Office
550 West 7th Ave., STE 1450
Anchorage, AK 99501

AMERICAN SAMOA
Forestry Program Manager
P.O. Box 5319 ASCC / AHNR
Pago Pago, AS 96799
011-684/699-1394
FAX 699-5011
e-mail: ssuemann@yahoo.com

ARIZONA
http://www.land.state.az.us/
Arizona State Land Department
2901 W. Pinnacle Peak Road
Phoenix, AZ 85027-1002

ARKANSAS
http://www.forestry.state.ar.us/
AR Forestry Commission
3821 West Roosevelt Rd.
Little Rock, AR 72204-6396

CALIFORNIA
http://www.fire.ca.gov/
Dept of Forestry & Fire Protection
P.O. Box 944246
1416 9th St., Rm. 1505
Sacramento, CA 94244-2460

COLORADO
http://www.colostate.edu/Depts/
  CSFS/
CO State Forest Service,
Colorado State University
203 Forestry Bldg.
Fort Collins, CO 80523

CONNECTICUT
http://www.dep.state.ct.us/
  burnatr/forestry/index.htm
Division of Forestry
79 Elm Street
Hartford, CT 06106

DELAWARE
http://www.state.de.us/deptagri/
    About_Us/forest.htm
Delaware Forest Service
2320 S. DuPont Highway
Dover, DE 19901

DISTRICT OF COLUMBIA
http://www.ddot.dc.gov/ufa/
    index.shtm
Urban Forestry Administration
District Dept. of Transportation
1105 O Street, SE
Washington, DC 20003

FLORIDA
http://www.fl-dof.com/
Division of Forestry
3125 Conner Blvd.
Tallahassee, FL 32399-1650

GEORGIA
http://www.gfc.state.ga.us/
GA Forestry Commission
P.O. Box 819
Macon, GA 31202-0819

GUAM
http://www.admin.gov.gu/doa/
    GOVGUAMID/
    DOAGR-ID_1.html
Territorial Forester
Forestry & Soil Resources Division
192 Dairy Road
Mangilao, Guam 96923

HAWAII
http://www.state.hi.us/dlnr/dofaw/
    index.html
Division of Forestry & Wildlife
1151 Punchbowl Street
Honolulu, HI 96813

IDAHO
http://www2.state.id.us/lands/
    index.htm
ID Department of Lands
954 West Jefferson St.
P.O. Box 83720
Boise, ID 83720-0050

ILLINOIS
http://dnr.state.il.us/conservation/
    forestry/
Division of Forest Resources
2005 Round Barn Road
Champagne, IL 61821

INDIANA
http://www.state.in.us/dnr/forestry/
Division of Forestry
Dept. of Natural Resources
402 W. Washington St.
Room W296
Indianapolis, IN 46204

IOWA
http://www.state.ia.us/forestry/
Department of Natural Resources
Wallace Office Building
East 9th & Grand Avenue
Des Moines, IA 50319

KANSAS
http://www.kansasforests.org/
Kansas Forest Service
2610 Claflin Road
Manhattan, KS 66502-2798

KENTUCKY
http://www.forestry.ky.gov/
KY Division of Forestry
627 Comanche Trail
Frankfort, KY 40601

LOUISIANA
http://www.ldaf.state.la.us/forestry/
    index.htm
Office of Forestry
P.O. Box 1628
Baton Rouge, LA 70821

MAINE
http://www.state.me.us/doc/mfs/
ME Forest Service
22 State House Station
Harlow Building
Augusta, ME 04333

REPUBLIC OF THE
MARSHALL ISLANDS
Ministry of Resources and
Development
Coconut St. (P.O. Box 1727)
Majuro, Republic of the
Marshall Islands 96960
011-692/625-3206
FAX 625-7471
e-mail: agridiv@ntamar.com

MARYLAND
http://www.dnr.state.md.us/forests
DNR - Forest Service
580 Taylor Avenue, E-1
Annapolis, MD 21401

MASSACHUSETTS
http://www.state.ma.us/dem/
    programs/forestry/index.htm
DEM P.O. Box 1433
Pittsfield, MA 01202

MICHIGAN
http://www.dnr.state.mi.us/
    SubIndex.asp?LinkID=168
    &sec=main&imageid=4
MI - DNR Forest Mgmt. Division
Mason Building., 8th Floor
Box 30452
Lansing, MI 48909-7952

FEDERATED STATES OF
MICRONESIA
http://www.fsmgov.org/info/
    natres.html
Dept. of Economic Affairs
PO Box PS-12
Palikir, Pohnpei FSM 96941

MINNESOTA
http://www.dnr.state.mn.us/
    forestry/
Division of Forestry
500 Lafayette Road
St. Paul, MN 55155-4044

MISSISSIPPI
http://www.mfc.state.ms.us/
MS Forestry Commission
301 N. Lamar Street, Suite 300
Jackson, MS 39201

MISSOURI
http://www.conservation.state.mo.
    us/
MO Dept. of Conservation
P.O. Box 180
Jefferson City, MO 65102

MONTANA
http://www.dnrc.state.mt.us/
    forestry/
DNRC - Forestry Division
2705 Spurgin Road
Missoula, MT 59804

NEBRASKA
http://www.ianr.unl.edu/nfs/
NE Forest Service
Rm. 103, Plant Industry Bldg.
Lincoln, NE 68583-0815

NEVADA
http://www.forestry.nv.gov/
Division of Forestry
1201 Johnson Street Suite D
Carson City, NV 89706-3048

NEW HAMPSHIRE
http://www.dred.state.nh.us/
    forlands/
Division of Forests & Lands
Box 1856 - 172 Pembroke Road
Concord, NH 03302-1856

NEW JERSEY
http://www.state.nj.us/dep/forestry
State Forestry Service
P.O. Box 404
Trenton, NJ 08625-0404

NEW MEXICO
http://www.emnrd.state.nm.us/
    forestry
Forestry Division
P.O. Box 1948
Santa Fe, NM 87504-1948

NEW YORK
http://www.dec.state.ny.us/
    website/dlf/index.html
NYS Dept. of Environmental Cons.
625 Broadway
Albany, NY 12233-4250

NORTH CAROLINA
http://www.dfr.state.nc.us/
NC Division of Forest Resources
1616 Mail Service Center
Raleigh, NC 27699

NORTH DAKOTA
http://www.ndsu.nodak.edu/ndsu/
    administration/forest
ND Forest Service
307 First Street
Bottineau, ND 58318-1100

NORTHERN MARIANA
ISLANDS
http://www.saipan.com/gov/
Territorial Forester
CNMI Dept. of Lands & Natural
Resources
P.O. Box 1007
Saipan, MP 96950

OHIO
http://www.hcs.ohio-state.edu/
    ODNR/Forestry.htm
Division of Forestry
1855 Fountain Square Ct., H-1
Columbus, OH 43224

OKLAHOMA
http://www.okag.state.ok.us/
   frt.htm
OK Dept. of Agriculture -
Forestry Services
PO Box 528804
Oklahoma City, OK 73152-3864

OREGON
http://www.odf.state.or.us/
OR Dept. of Forestry
2600 State Street
Salem, OR 97310

PALAU
Chief Forester
Palau Agriculture and Forestry
Erenguul St. (P.O. Box 460)
Koror, Palau 96940
011-680/488-2504
FAX 488-1475
DAMR@Palaunet.com

PENNSYLVANIA
http://www.dcnr.state.pa.us/forestry
Bureau of Forestry
P.O. Box 8552
Harrisburg, PA 17105-8552

PUERTO RICO
Forest Service Bureau - DNER
P.O. Box 9066600, Puerta de Tierra
San Juan, PR 00906-6600
787/725-9593
FAX 721-5984
e-mail: manriveortiz@hotmail.com

RHODE ISLAND
http://www.state.ri.us/dem/
   programs/bnatres/forest/
   index.htm
Div. of Forest Environment
1037 Hartford Pike
North Scituate, RI 02857

SOUTH CAROLINA
http://www.state.sc.us/forest/
SC Forestry Commission
P.O. Box 21707
Columbia, SC 29221

SOUTH DAKOTA
http://www.state.sd.us/doa/forestry/
   index2.htm
Resource Conservation & Forestry
Foss Building
523 E. Capitol Ave.
Pierre, SD 57501

TENNESSEE
http://www.state.tn.us/agriculture/
   forestry/
TN Dept. of Agriculture -
Division of Forestry
P.O. Box 40627
Melrose Station
Nashville, TN 37204

TEXAS
http://txforestservice.tamu.edu/
Texas Forest Service
301 Tarrow Dr., Suite 364
College Station, TX 77840-7896

UTAH
http://www.nr.utah.gov/slf/
   slfhome.htm
Dept. Natural Resources
1594 W. North Temple Suite 3520
Salt Lake City, UT 84114-5703

VERMONT
http://www.state.vt.us/anr/fpr/
   index.htm
Dept. of Forests, Parks and
Recreation
103 S. Main Street
Waterbury, VT 05671-0601

VIRGIN ISLANDS
http://www.usvi.org/agriculture/
Commissioner of Dept. of
Agriculture
Estate Lower Love - Kings Hill
St. Croix, US VI 00850

VIRGINIA
http://state.vipnet.org/dof/
   index.html
VA Dept. of Forestry
900 Natural Resources Drive,
STE 800
Charlottesville, VA 22903

WASHINGTON
http://www.wa.gov/dnr
Dept. of Natural Resources
Box 47001
1111 Washington Street
Olympia, WA 98504-7001

WISCONSIN
http://www.dnr.state.wi.us/org/
   land/forestry/
DNR - Division of Forestry
P.O. Box 7921
Madison, WI 53707

WEST VIRGINIA
http://www.wvforestry.com/
WV Division of Forestry
1900 Kanawha Blvd., East
Charleston, WV 25305-0180

WYOMING
http://lands.state.wy.us/
WY State Forestry Division
1100 West 22nd Street
Cheyenne, WY 82002

NASF Washington Office
http://www.stateforesters.org/
Hall of the States
444 North Capitol Street
Suite 540
Washington, DC 20001
202/624-5415
FAX 624-5407
e-mail: nasf@sso.org

EVACUATION CHECKLISTS

The Evacuation Checklists discussed in Chapter 2: Preparation are reproduced on the next several pages for your convenience. You may photocopy these checklists, or create your own checklists using these as a guide.

The "Emergency Telephone List"

Keep a list of emergency telephone numbers in a conspicuous place in your home and in all of your vehicles. It is even a good idea to make additional copies of the list so that family members and visitors will each have one. Most cell phones can be programmed to store dozens, if not hundreds, of names and numbers, so take advantage of that feature if you have it. Your list might include the following numbers:

☐ 911, and instructions on when it is appropriate to use it.

☐ All responding fire departments in your area (there may be more than one).

_____

_____

☐ Local law enforcement and emergency medical dispatchers.

_____

_____

☐ The state patrol number for information on road conditions and possible closures.

_____

☐ Your children's schools and daycare facilities.

_____

_____

☐ People who are authorized to pick up your children if you are unable to.

_____

_____

☐ Relatives, close friends, and others outside the potential fire zone who should be notified of your whereabouts.

_____

_____

☐ Your neighbors.

_____

_____

☐ Your business office or employer.

_____

_____

☐ Your physicians, pharmacies, veterinarians, and medical facilities.

_____

_____

☐ Boarding facilities for horses and other large domestic animals.

_____

_____

☐ Other:

_____

_____

_____

_____

_____

The "Last-Minute Grab List"

These items are things you use or need to have available on a day-to-day basis, so it doesn't make sense to pack them up weeks in advance of a possible evacuation. However, you should consider keeping this list and a large plastic storage container (evacuation box) by your door or perhaps in your front hall closet. The list will remind you of everything you need to grab; the box provides a convenient receptacle into which you can toss everything quickly and carry it out with you as you leave. Your Last-Minute Grab List might include the following items:

☐ All *medications* and medical supplies for everyone in your family and your pets.

☐ Your driver's license and passport.

☐ Any cash, checkbooks, credit and gift cards, calling cards, and similar items.

☐ Your address book, day planner, or PDA, and a copy of your Emergency Telephone List.

☐ Portable electronic devices, such as your cell phone and charger, and your laptop computer and peripherals. If you have backed up your home computer files and operating system, take those disks, too.

☐ Glasses and contact lenses, with cleaning and storing supplies.

☐ A special blanket or stuffed animal that your child will want at bedtime.

☐ The children's school textbooks and notebooks.

☐ Jewelry that can't be replaced (take only the "real" stuff).

☐ Cameras, exposed but undeveloped film, and video equipment.

☐ Pet stuff, such as leashes, medicine, and food.

☐ _____

☐ _____

☐ _____

☐ _____

The "Pre-Packed Items List"

Your actual choices will vary according to where you'll stay or how long you'll be gone when you evacuate, but the following Pre-Packed Items List is a good start on planning for the things you'll need. Give some thought to what additional items should be on your list, according to your particular needs. It's a good idea to have all of your pre-packed items boxed up and stacked in a closet by the door or next to your car, preferably in easy-to-carry plastic storage containers (evacuation boxes) with lids. Some people even spend the whole fire season with all these items in their cars.

☐ This book! (Chapter 4: Recovery contains useful information on what to do after the fire.)

☐ Copies of all prescriptions and medical records for you family, including a list of all prescription numbers, where they are on file, and the pharmacy phone numbers, especially those obtained by mail-order.

☐ Additional copies of your Emergency Telephone List.

☐ Insurance information, including your policy numbers, agents' names and phone numbers, and insurance cards.

☐ Photocopies of all the documents in your safe-deposit box.

☐ Books and toys for the children.

☐ Several days' worth of clothing and a jacket or sweater for each person.

☐ Toiletries and other personal care items.

☐ Non-perishable snacks and bottled water.

☐ Pet carriers, crates, toys, extra food, and feeding dishes, as well as feed, tack and accessories, grooming supplies, and medications for horses and other large domestic animals.

☐ Copies of your children's school records, scrapbooks, and any collections of school projects, awards, and achievements you've been saving for them.

☐ Other items you can easily carry and would hate to lose, such as family photos, mementos, and heirlooms, as well as home movies or video tapes of children's birthdays, graduations, weddings, and other family events.

☐ _____

☐ _____

☐ _____

☐ _____

The "To Do Before Leaving List"

The fourth evacuation list is the "To Do Before Leaving List." With several hours' notice, you may have time to accomplish a number of additional tasks that will help to protect your home and make the firefighters' jobs easier. You may want to copy this page from the book or make your own list and add the things that are specific to your situation and keep it with your other lists as a reminder.

- ☐ Shut off the gas at the meter or propane tank.

- ☐ Leave your outside (and inside) lights on to help firefighters find your house through smoke and darkness. This also lets firefighters know there is power to the house; this means your well or water system should still be working.

- ☐ Close windows, shutters, vents, fire resistant or heavy drapes, blinds, and doors to help block radiant heat. Locking the doors and windows is up to you. Leaving them unlocked gives firefighters easy access without having to break in, but obviously it also leaves your house unsecured.

- ☐ Take down sheer curtains; they are very flammable.

- ☐ Open the fireplace damper to allow hot air to vent out of the house. Your chimney should already have a spark arrestor installed, but as an added precaution, position the fireplace screen over the hearth to prevent sparks and embers from being blown in.

- ☐ Move furniture to the center of each room. Try to get flammable items away from doors and windows, especially large glass doors and windows.

- ☐ Wet down any plants or trees within 15-20 feet of your house and other buildings.

- ☐ Put a sprinkler on your roof (unless it's composed of non-combustible materials, such as tile or metal). Wet the roof down before you leave.

- ☐ Put garden hoses and buckets filled with water around the house. Firefighters can use these to put out spot fires and flare-ups, if necessary.

- ☐ Position a metal ladder next to your house for firefighters to use to access your roof.

☐ Seal outside vents with heavy plywood or non-flammable barriers, such as rocks and gravel.

☐ _____

☐ _____

☐ _____

☐ _____

INDEX

References are to page numbers.

Notes

Notes

Notes

Notes

Notes

Notes

Notes

Notes

ABOUT BRADFORD PUBLISHING

Founded in 1881, Bradford Publishing Company is Colorado's oldest and most trusted publisher of legal forms and information. Today Bradford Publishing has an inventory of more than 800 legal forms and books specific to Colorado law. Our forms are accurate and up-to-date because we consult with attorneys and state agencies to keep them that way. All of our products are available on our website and many of our forms can be downloaded and completed on-screen. Bradford forms are accepted by the Colorado courts.

Visit the books section of our website to see our growing list of legal publications. If you live or work in the Denver Metro area, you can find all our books, forms and supplies at our store in lower downtown.

BRADFORD PUBLISHING COMPANY

1743 Wazee Street

Denver, Colorado 80202

800-446-2831

303-292-2590

303-298-4014 Fax

For quantity discounts on *Living With Wildfires*, please give us a call.

www.bradfordpublishing.com